Advances in Structural Health Monitoring

Edited by Maguid H.M. Hassan

Published in London, United Kingdom

IntechOpen

Supporting open minds since 2005

Advances in Structural Health Monitoring
http://dx.doi.org/10.5772/intechopen.80933
Edited by Maguid H.M. Hassan

Contributors
Ravibabu Mulaveesala, Geetika Dua, Vanita Arora, Kodai Matsuoka, Tsutomu Watanabe, Tat-Hean Gan, Shehan Lowe, Anurag Dhutti, Mostapha Tarfaoui, Khalid Lafdi, Yumna Qureshi, Khalil K. Lafdi, Rui Pitarma, João Crisóstomo, Yoann Hebrard, Maguid H.M. Hassan

First published in London, United Kingdom, 2019 by IntechOpen
IntechOpen is the global imprint of INTECHOPEN LIMITED, registered in England and Wales, registration number: 11086078, 7th floor, 10 Lower Thames Street, London, EC3R 6AF, United Kingdom
Printed in Croatia

British Library Cataloguing-in-Publication Data
A catalogue record for this book is available from the British Library

Additional hard and PDF copies can be obtained from orders@intechopen.com

Advances in Structural Health Monitoring
Edited by Maguid H.M. Hassan
p. cm.
Print ISBN 978-1-83962-872-6
Online ISBN 978-1-83962-873-3
eBook (PDF) ISBN 978-1-83962-874-0

For EU product safety concerns:
IN TECH d.o.o., Prolaz Marije Krucifikse Kozulić 3, 51000 Rijeka, Croatia, info@intechopen.com or visit our website at intechopen.com.

Meet the editor

Professor Maguid H.M. Hassan earned his PhD in Structural Reliability from the University of Maryland at College Park, USA. He participated in the establishment of Civil Engineering programs at several private higher education establishments in Egypt. He was awarded a Fulbright teaching/research grant in 2004 where he taught structural engineering courses at the University of Maryland at College Park for one academic year and conducted his research. Prof. Hassan joined the British University in Egypt (BUE) in 2006 to establish the Civil Engineering Department. In September 2013, Prof. Hassan was appointed Associate Dean for Research, Post Graduate Studies and Community Services. In November 2014 he was appointed Dean of Engineering. Prof. Hassan's research interests include smart structural systems, structural health monitoring, structural reliability, structural steel analysis and design, applications of fuzzy logic and neural networks and smart structural control.

Contents

Preface XI

Chapter 1 1
Introductory Chapter: Advances in Structural Health Monitoring
by Maguid H.M. Hassan

Chapter 2 5
Structural Health Monitoring from Sensing to Processing
by Yoann Hebrard

Chapter 3 23
The Importance of Emissivity on Monitoring and Conservation of Wooden Structures Using Infrared Thermography
by João Crisóstomo and Rui Pitarma

Chapter 4 39
Applications of Infrared Thermography for Non-destructive Characterization of Concrete Structures
by Ravibabu Mulaveesala, Geetika Dua and Vanita Arora

Chapter 5 55
Monitoring of Critical Metallic Assets in Oil and Gas Industry Using Ultrasonic Guided Waves
by Anurag Dhutti, Shehan Lowe and Tat-Hean Gan

Chapter 6 75
Application of a Frequency-Based Detection Method for Evaluating Damaged Concrete Sleepers
by Kodai Matsuoka and Tsutomu Watanabe

Chapter 7 97
Nanotechnology and Development of Strain Sensor for Damage Detection
by Yumna Qureshi, Mostapha Tarfaoui, Khalil K. Lafdi and Khalid Lafdi

Preface

Structural health monitoring has emerged as a viable tool for damage detection and preventive maintenance procedures for engineered structures. This book covers themes such as sustainable design, smart structural health monitoring, conservation of historic cultural heritage, advanced damage assessment techniques and reliability of monitored engineering systems. These themes are considered to augment conventional structural health monitoring with the objective of achieving a sustainable system design.

The book opens with a description of a new sensor design proposed for the aeronautics industry. This sensor is expected to enable maintenance operators to determine the load level seen by an engine links after hard landings. Infrared thermography is presented as a viable non-destructive tool in monitoring wooden cultural heritage structures, which is characterized as a friendly, noninvasive method that suits the nature of cultural heritage structures. The same technology is also proposed for non-destructive monitoring of concrete structures where a novel aperiodic thermal wave imaging technique is employed for testing and evaluating rebar corrosion in concrete structures. Ultrasonic guided waves technology is also proposed for metallic structures to support their integrity and maintenance management. Applications in pipelines and storage tanks, which are critical assets in the oil and gas industry, are considered. Frequency-based damage detection is considered for evaluating damaged concrete sleepers and its practical use is demonstrated through laboratory and field tests. Finally, a new generation of strain sensors in the form of wire/thread that can be incorporated into composite materials to detect damage are presented. The proposed microscale strain sensor consists of flexible, untwisted nylon yarn coated with a thin layer of silver using an electro-less plating process. All such applications are meant to introduce new trends and/or new applications of new and/or existing structural health monitoring techniques.

The editor would like to thank all contributors whose work is represented in this book and whose influences are expected to help in shaping the future of structural health monitoring.

Maguid H.M. Hassan
Dean of Engineering,
The British University in Egypt (BUE)

Chapter 1

Introductory Chapter: Advances in Structural Health Monitoring

Maguid H.M. Hassan

1. Introduction

Structural health monitoring has emerged as a viable tool for damage detection and preventive maintenance procedures. There is a wide range of proposed applications that were documented in the literature over the past two decades. Recently, the notion of sustainable design has emerged as an important attribute of all engineering designs. Sustainable design is defined as one that would result in systems that are smart, optimum, and reliable.

In this book, the concept of sustainable design is the backbone of the design of a structural health monitoring system. Within the backdrop of the presented definition, this book attempts to present structural health monitoring as a tool that would result in a sustainable engineering system. There are several aspects that are now being introduced to the conventional notion of structural health monitoring which are expected to contribute to such objective. Smart systems, smart materials, wireless sensor networks, conservation of historic cultural heritage, and autonomous systems are some of these aspects.

The book explores the design of smart structural health monitoring systems, using smart technologies and/or materials. It presents the issue of conservation of heritage structures using structural health monitoring tools. It explores the optimum employment of sensor networks that would render the most optimum structural health monitoring system. This book attempts to present such advanced concepts and/or technologies as the new direction of structural health monitoring.

This chapter briefly presents several advanced observations and/or applications that are considered to augment structural health monitoring techniques currently in practice.

2. Sustainable design

Sustainable design is now considered an essential requirement for all engineering systems. Sustainability is introduced as a general feature that reflects a set of objectives that should be achieved within the designed system. A sustainable system is expected to be optimum, smart, reliable, and recyclable. Initially, sustainability was always related to recyclable materials that are used in constructing engineering systems. With the evolution of smart materials, the introduction of smart systems, and the innovation associated with the internet of things, sustainability should now extend to include a much wider set of objectives, as outlined above.

Structural health monitoring, which is designed to provide a mechanism for damage detection and preventive maintenance strategies, contributes to most of this set of objectives in order to attain sustainable engineering systems. Integrating structural health monitoring within engineering systems results in a smart, reliable, and optimum system, smart in the sense that the system can detect damaged components, independently, and can propose recommendations to safeguard against any potential failures; reliable in the sense that such early detection of damage and the resulting recommendations of preventive maintenance activities would directly improve the reliability of engineering systems; and finally, optimum in the sense that such integrated smart systems would result in designs that ultimately would require less materials, less maintenance, and less failure occurrences.

3. Smart structural health monitoring

Structural health monitoring has gained an increased interest and research activity over the past two decades. By definition structural health monitoring comprises four main tasks, namely, damage identification, damage localization, damage severity evaluation, and structural system life expectancy prognosis. All current research activities are concentrating on the first two tasks, i.e., damage identification and localization. The continuous developments and evolutions in the applications of smart materials and technologies lend themselves to ample applications in the development of smart structural health monitoring systems.

Smart structural health monitoring systems are the ones that employ smart materials in designing their sensor networks and/or smart technologies in designing their diagnostic and inference systems. In light of the broad definition of sustainable design, presented earlier, such a smart health monitoring system, when integrated with any given structural system, would result in a sustainable engineering system.

4. Conservation of historic cultural heritage

Historic cultural heritage structures formulate the collective identity of a nation. It is the responsibility of current generations to ensure their safety and integrity in order to be safely maintained for future generations. This includes any conservation and/or retrofitting activities that might be required. It is expected that such structures would pose a unique challenge in that regard due to their age and the extreme environmental conditions they faced over the years. Another significant challenge is the materials used in their construction and the structural systems which in general would be inconsistent with current practices.

The introduction of structural health monitoring, as a conservation tool for such structures, is a state-of-the-art concept that would address most of the identified challenges, presented above. Structural health monitoring systems could be designed to integrate with such historic structures, thus, providing real-time monitoring mechanisms that can detect any damages, as they develop. Such early identification would ensure the timely introduction of maintenance activities that would ensure the safety of the historic structure.

5. Advanced damage assessment techniques

Damage identification is the major function of any structural health monitoring system. There is a wide range of damage assessment methods and techniques that have been employed in the published health monitoring applications. With the development of smart materials and advanced technologies, new damage assessment techniques are introduced every day. Some of these techniques could be successful in certain areas; however, integrating them with structural health monitoring is the new advancement.

Most of the advanced damage assessment techniques fall within the nondestructive evaluation techniques. Some of these advanced techniques are acoustic emission, infrared thermography, ultrasonic guided waves, and wavelet transforms.

The introduction of composite materials and their increasing employment in building engineering systems led to the development of a new breed of damage assessment techniques using microscale strain sensors that are embedded within composite materials.

6. Reliability of monitored engineering systems

The reliability of any designed engineering system is a major concern. Due to the uncertainty inherent in several types of loads and the randomness in material properties, complete safety is not possible to attain. There is always a certain probability of failure that is associated with any designed engineering system. It is only important to ensure that such probability of failure is acceptable in terms of the nature of application at hand. Furthermore, it is important to be able to evaluate such probability and thus relate that to the reliability of any designed system.

The integration of structural health monitoring, within an engineering system, is expected to enhance its reliability. It is important to consider the reliability, of such system, in two main levels, the first is the reliability of the structural health monitoring system itself and the second is the reliability of the monitored engineering system as a whole, including the health monitoring enhancement. There is not enough research activity in the area of reliability of structural health monitoring systems and their impact on the reliability of monitored systems. This requires a targeted research effort in this area which would support the practical implementation of such systems, especially when dealing with the conservation of historic cultural heritage.

7. Summary

The previously presented concepts are some of the new advancements in the design of structural health monitoring systems. It is envisaged that such advancements would lead to smart structural health monitoring that would result in sustainable engineering systems. Sustainability is considered an important objective in today's engineering design that is realized due to the current state of climate change and global warming, which are hugely aggravated by industrial and construction activities. It is intended through the chapters of this book to present demonstrations of and applications for such new advancements in order to encourage further research and implementation of such advanced techniques and technologies in structural health monitoring.

Author details

Maguid H.M. Hassan
The British University in Egypt (BUE), Cairo, Egypt

*Address all correspondence to: mhassan@bue.edu.eg

Chapter 2

Structural Health Monitoring from Sensing to Processing

Yoann Hebrard

Abstract

Providing the best availability of aircrafts is a key driver in aeronautics industry. Monitoring system able to detect signs of failure before they happen, thanks to sensors and diagnosis/prognosis algorithms, is key for improving aircraft operability. Since a suspension system is connecting the engine to the aircraft, after hard landing, aircraft companies need to know if the suspension system is safe or could have been damaged. This chapter presents an autonomous wireless load sensing recorder development that will enable maintenance operators to make a relevant diagnosis of the suspension system by measuring the load level seen after a hard landing by connecting a portable device near the embedded sensor system. The sensor integrates energy harvesting and RFID communication modules that have been developed for this application. Data acquisition is performed by an embedded microcontroller connected to sensors. The paper is firstly dedicated to the different energy sources available in the project application (engine pods). The second part gives a presentation of the various devices developed for converting ambient energy into electric power and SHM system. The last part presents real measurement of ambient energy level from real tests in comparison to the energy needed to power the system.

Keywords: SHM, wireless sensor, RFID, energy harvesting, engine health management

1. Introduction

Based on nondestructive testing (NDT) technologies, systems known as structural health monitoring (SHM) make it possible to anticipate the deteriorations of a structure to avoid accidents. Associated with statistical processing systems, they also make it possible to optimize product life while reducing maintenance costs.

The technologies of structural health monitoring, or SHM, meet the requirement of maintaining the quality of products over time. Their main objective is to ensure the health of structures, extend their life, anticipate their failures, and enhance their performance. The SHM is therefore for the industry part of a strategy link to economical, commercial, and safety drivers.

The SHM approach is indeed both a set of processes and a control strategy [1–3]. It consists of monitoring continuously or at regular intervals the integrity of a structure by detecting cracks or alterations, such as delamination of composite materials. The proposed solutions generally include sensor networks, for example, piezoelectric or optical fiber type, coupled to signal processing systems.

The main contributions of the SHM with respect to the NDT are the integration and automation of control in a global strategy of control and/or maintenance of structures. The SHM makes it possible to improve knowledge of the structures by better monitoring, reducing maintenance interventions, and optimizing the materials used.

The ultimate goal of the SHM is the development of autonomous, continuous monitoring systems capable of detecting structure damage in real time to avoid accidents. The key challenges are in the early detection of damage, allowing optimal maintenance.

Several factors have contributed to the development of the SHM approach in recent years: numerical simulation capabilities and algorithms and the reduction of sensor consumption and its wireless communication.

This chapter presents an autonomous wireless load recorder development to enable maintenance operators to get access to loading histogram seen by an engine link in service by simply setting a portable reading device near the attachment system. Energy harvesting and RFID communication modules have been developed for this sensor considering application data such as flight duration, environmental parameters, and theoretical loading cases. Sensor system includes embedded sensors connected to a microcontroller for data acquisition and wireless communication management and energy harvesting modules. RFID technologies are used for communication and to power the device when the engine is not running. Newer SHM systems are demanding low-power sensors and wireless communication system to enable data acquisition and diagnostic automatization. Thus, several researchers have recently investigated such techniques to extract wirelessly communicate data stored in these stand-alone systems. The first part of the paper makes a comparison between different wireless communication protocols against RFID on the market considering application needs. The second part presents the RFID system developed for transmitting local data from the embedded memory of the system for analysis in a SHM diagnostic module. The last part presents an evaluation test of system communication capabilities and provides a maturity assessment on the technology.

2. RFID sensor concept and constraints for SHM application

Radio-frequency identification (RFID) uses electromagnetic fields to automatically identify and track tags attached to objects. The tags contain electronically stored information. The information can be used for different purposes. One purpose is traceability of engine parts. Today, the configuration of the engines in service must be known to respect the legal constraints. Its follow-up is therefore the responsibility of its operator. Companies realize the traceability of parts on software package in order to manage the maintenance planning, carry out the use of particular parts (life-limited parts, for example), and give a monthly status of reliability to the authority. But there is some limitation about manual software tracking. The configuration and the usage of LRU parts may not be robustly documented in the tool. In fact, the modifications carried out as part of the line maintenance can, for practical reasons, be saved later on the maintenance software. So, these data are not always up to date in the database.

The "RFID" solution seems particularly well positioned to meet this need. Its design was motivated to meet the high constraints on the configuration management at the aircraft and assembly level. It allows to know the reference of the part in a precise, fast, and automatic way. In addition, the standardization of the data in the memory of RFID chip allows a great interoperability between integrating suppliers, aircraft manufacturers, airlines, vendors, and repair shops.

But this traceability is not only related to configuration management. It can also be applied to get the real conditions of the part. RFID can be used as a sensor to gather information about real-world conditions. Alerts can be launched when some constraints have been exceeded during the usage of the parts. Visual inspection can be replaced by RFID inspection. Each of the parts keeps information about its own health..

This makes it possible to:

- Identify the MTBF specific to the use of each part (number of cycles, average temperature, number of hours in operation, etc.).
- Estimate the remaining useful life which will make it possible to value each piece individually by bringing to it a traceability of its usage.
- Follow the end of each part's life cycle by alerting the reuse of discarded parts as well as to alert in the event of counterfeit use copying a serial number already installed or unknown to manufacturers' databases.
- Minimize the risk of mounting errors, by providing tools that will check the compatibility between several parts with RFID tag.
- Give all information about a given part (manufacturing data or "birth record," inspection instructions, assembly/disassembly, troubleshooting, etc.).

RFID tag has been identified as the missing link for the realization of Integrated Engine Health Management.

But for being used in aeronautics, the RFID tag must meet different criteria and constraints. They must respect extreme environmental conditions (from −65 to +150°C). They need to comply with applicable legislation and have no impact on existing systems and equipment. The usage of RFID transmission is limited to ground operations (no application on taxiways or runways or aircraft in motion). For radio-frequency, no backscatter above 35 dBuv/m and no overflow on frequency bands are allocated for other uses (including for harmonics). RFID readers must comply with GS1/EPC class 1 Gen 2 (ISO180000-63) standards operating on the 860–960 MHz compatible frequency range around the globe.

2.1 RFID sensor architecture

RFID systems use one or more RFID interrogators/readers and several passive RFID tags. For communication with the tag, the reader emits a high-frequency modulated electromagnetic wave. The tag receives the information modulated in the field, and it can answer to the request from the reader with a backscatter operation, i.e., the reflection coefficient is changed with respect to the requested information (**Figure 1**). In addition to the transmission of information between the

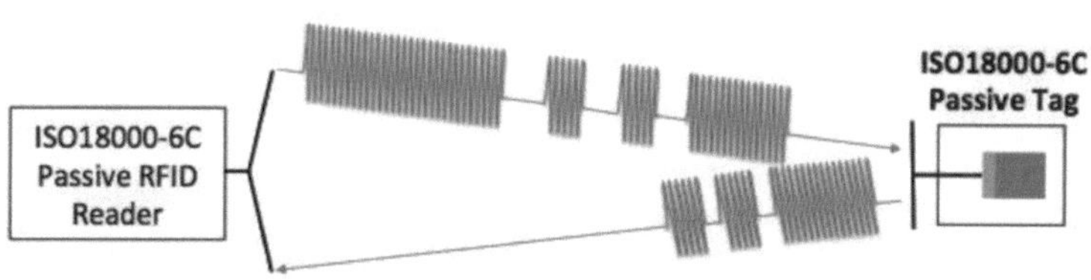

Figure 1.
RFID backscattering.

tag and the reader, the tag can receive the necessary energy for operation from the electromagnetic field of the reader.

Due to regulatory and physical constraints, the used frequency band of the passive RFID tag influences the available data rate and sets an upper limit for its operation range. HF-RFID tags operate in the near field of the reader antenna, and so the operational distances are limited between several centimeters up to 1 m. By comparison with HF-RFID systems, UHF-RFID tags transmit their data and energy by the means of the electrical far field and thus offer a higher read range from 3 up to 6 m.

The RFID sensor tag is composed of (**Figure 2**).

A sensor interface converts the change of value of a sensor into something that can be properly treated. The sensor interface is composed of a sensor readout circuitry (charge amplifier, resistive bridge, etc.) and an analogue to digital converter (ADC). The sensor interface can either be passive (the readout electronics and the analogue to digital converter are fully powered by the reader's field) or semi-passive: an additional battery powers up the interface as well as the logic.

The signal interface adapts the external signals (sensor reading, data logging, microcontrollers, display, keyboard, etc.) to the standardized RFID tag. The signal interface can be of several natures, like a serial bus interface such as SPI and I2C to connect directly the logical part of the RFID tag to an additional block such as a data logger, microcontroller, display, etc. In this case, the RFID sensor tag can be either semi-passive or active.

A logic part is the translator between the front end and the sensor interface by coding, decoding, commanding, processing, and storing information. The logic implementation usually follows a defined standard and a certain associated protocol.

An analogue RF front end typically contains a rectifier circuitry to convert RF power into DC, a clock, a modulator, and a demodulator.

An antenna is directly matched to the tag's front-end impedance to communicate with the reader. Different solutions exist for antennas, the RF front end and the logical part which are treated in the literature, whereas a low-power signal interface is a relatively new concept in the RFID context. In the case of sensor reading, the mixed signal interface should implement low-power architecture for passive, semi-passive, and even for active solutions for life span purposes. Solutions for low-power architectures are detailed in the following paragraphs.

3. Project CMOS-based RFID sensor

This PCB-based component is a passive RFID tag with embedded MCU (microcontroller unit) to manage communication protocol and application. RFID solution is a passive radio system without emission. The architecture is presented in **Figure 3**.

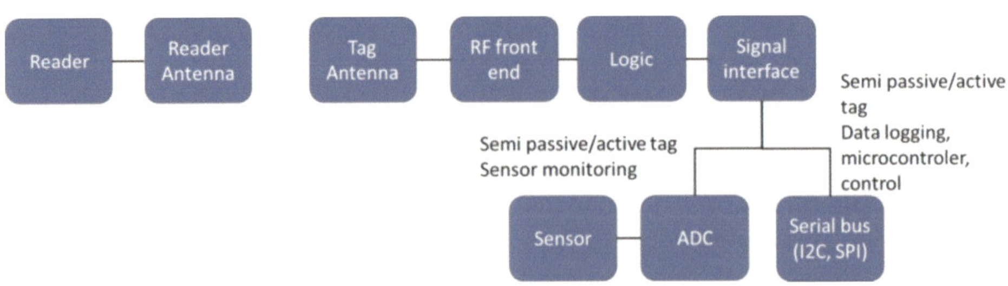

Figure 2.
RFID sensor architecture.

DOI: http://dx.doi.org/10.5772/intechopen.86758

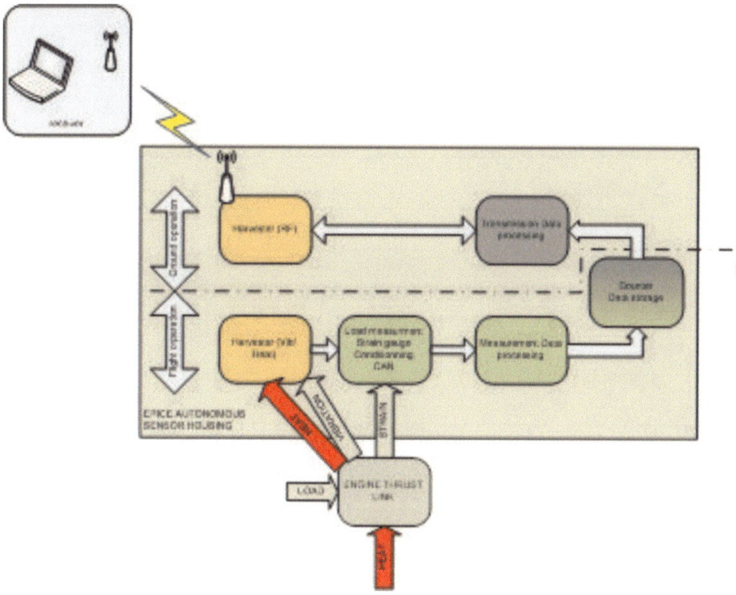

Figure 3.
RFID sensor architecture.

From a PCB point of view, a modular approach is proposed based on a motherboard including the existing RFID tag design and harvesting boards specific to each energy source.

The device architecture includes the following subsystems:

- A strain gauge bridge. The bridge measures the load supported by the engine link. The sensor signal is stored in the NVM using the threshold crossing detection techniques. This approach reduces the amount of data to be stored since a histogram of loading class will be provided between each monitoring cycles.

- Energy harvesters to supply embedded DC converters for strain gauge bridge and the nonvolatile memory power needs. Energy is harvested from vibrations and thermal differences of the engine.

- Data and power management done by a microcontroller. Data are stored into a NVM that will be interrogated using a wireless RFID transmission. Energy harvesters provide energy, but it must be stored and managed to supply strain gauge and memory. All these parts are managed by an integrated electronic device.

- An interrogating system. It is a RFID interrogator and is not embedded into the road but approached by an operator.

Circuits for converting piezoelectric material strain into useable electrical energy have been previously described (see [4–7] for details).

The module targeted performances are the following (**Table 1**):

System power needs have been established in order to size the different harvesting modules. **Figure 4** displays voltage and current measurements done at the power input level of the system. Several commands sent at the microcontroller level allow to identify associated consumption. Cycle is quite complex; however

Subsystem	Parameters	Value
	Ambient temperature	−40/150°C max
Strain gauge acquisition	Deformation range	±2500 µdef
	Measuring bandwidth	100 Hz
	Gain/offset	Configurable
	Max sampling frequency	1000 Hz
	Resolution	12 bits
RFID UHF	Read range	15 cm/1 m in free space
	Standard	EPC GEN 2 (EU frequencies)
Vibration harvester	Frequency of ambient vibration (maximum harmonic)/ acceleration amplitude	80–110 Hz/1 g
	Variability of the vibration frequency/acceleration amplitude	Random
	Weight/volume	60 g/30 cm^3
Thermal harvester	Hot/cold temperature	80/100°C
	Heat cycle	Variable
	Weight/volume	20 g/5 cm^3

Table 1.
System specification.

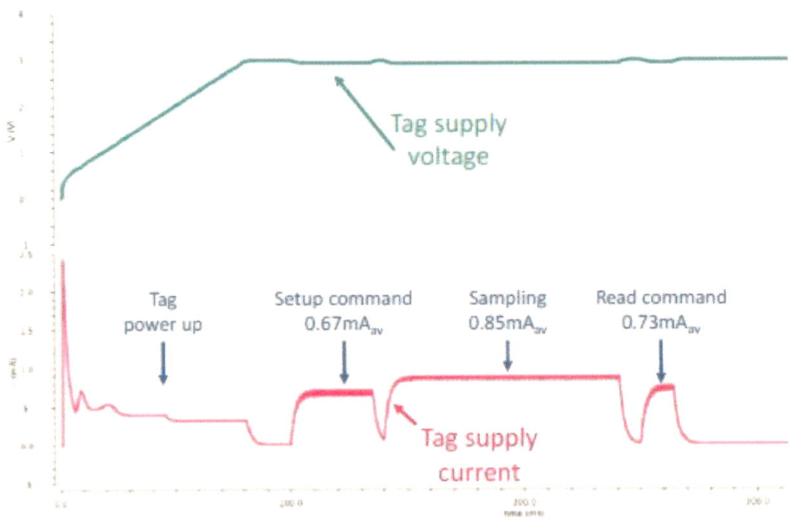

Figure 4.
System consumption identification in temperature.

a power need of 3 mW has been identified for the system to work according measurement needs.

3.1 Design of key elements

Key elements for piezoelectric energy conversion include impedance matching circuits and an energy management circuit which monitor the voltage levels.

Data and power are managed thanks to the following elements:

- RFID tag: it is a small radio device also called transponder or smart tag (one can refer to ref. [8] for general details about RFID technology). The tag includes a microchip managing RF communications and a flat aerial. System tag works in the ultrahigh frequency (UHF) band (850–950 MHz) as this band offers the best compromise in power consumption with a high range ~10 m in free environment. (Details on RFID frequencies can be found in [9]).

- An electronic platform (microcontroller, volatile memory, power management circuit, strain gauge conditioning circuit): this platform locally counts the strain threshold crossings and stores them into a nonvolatile memory. Strain gauge resistance value is 5000 Ohms, to reduce the power needed to energize the bridge.

Electronic platform can be energized in two ways depending on the operational phase:

- When application is on: energy is provided by energy harvesting modules.

- When application is off: equipment does not need to be monitored, but it does when the operator downloads load counter data. Energy harvester cannot supply energy for the NVM or the electronic platform. But RFID technology gives the possibility to harvest energy from the interrogating radio-frequency wave. When engine is off, energy is provided through the RFID link.

- Reader or interrogator: it sends and receives radio-frequency data to and from the tag using antennas. A reader may have several antennas that send and receive radio waves.

3.2 Available energy sources for energy harvesting

A multitude of energy sources are available in aircraft that can be accessed with energy harvesting technologies: temperature gradient and variation, vibrations, strain, ambient light, pressure changes, electrostatic charges, etc. (**Figure 5**).

However, not all sources have sufficient potential to provide enough energy to power a sensor system. The most critical parameter for comparing these technologies in the scope of aircraft applicability is their power-to-weight ratio (per flight cycle).

Reliability of these devices is also a key criterion. Among all potential sources, thermal and vibrational energies seem most likely to meet the sector constraints [4].

3.3 Vibration harvester

The operating principle of a piezoelectric generator is relying on direct piezoelectric conversion. The internal generation of electrical charge results from an applied mechanical load. Two families of piezoelectric micro-generator can be found. The first one, named direct coupling, is directly bonded to the host structure (**Figure 6a**). The second, called indirect or seismic coupling, is connected to the host structure through a secondary element, e.g., a clamped beam (**Figure 6b**).

The project used a piezoelectric composite material (PZT type P2) manufactured as beam coupled to the mechanical trust in seismic mode. This flexible composite is made of many PZT fibers and is highly damage tolerant and fatigue resistant.

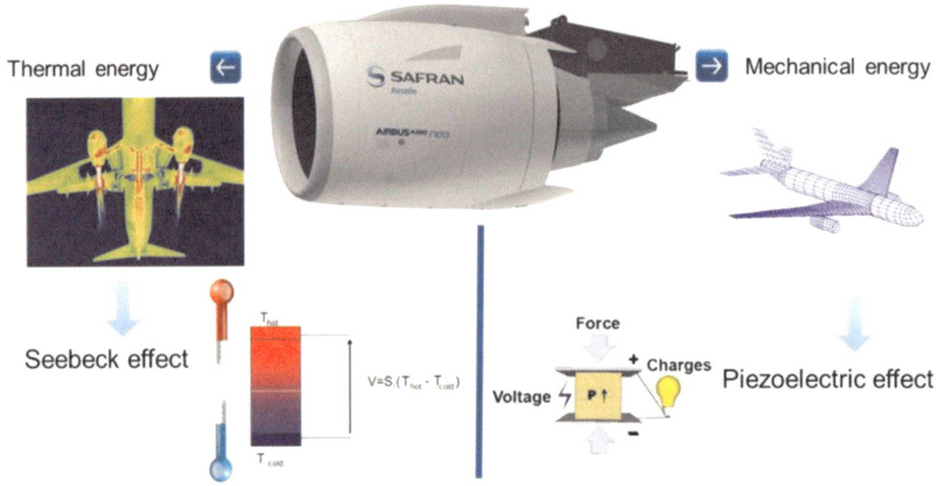

Figure 5.
Typical localizations of energy sources in engine environment.

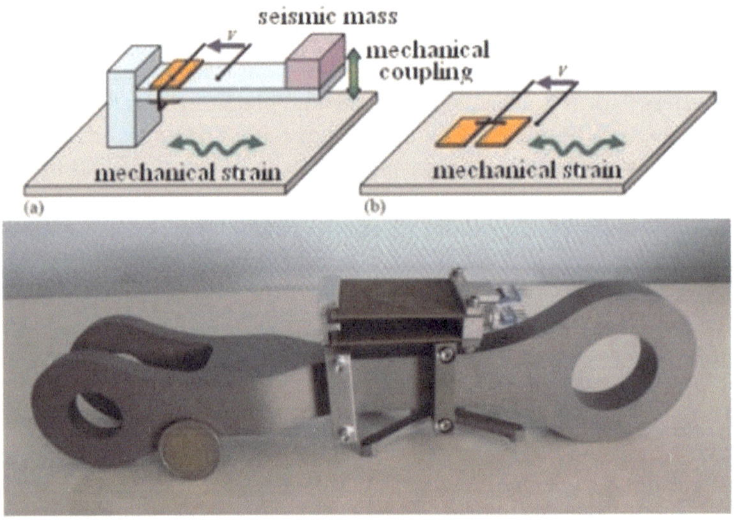

Figure 6.
Piezoelectric coupling: (a) seismic and (b) directly applied to real harvester.

The piezoelectric beams provide substantial power densities (250 µW/cm^3 at 120 Hz and 2.5 m/s^2) that make them particularly attractive for various energy harvesting applications. Seismic coupling had been chosen for the vibration harvester, because it can provide up to 10 times more energy than the classical method.

To keep a stable voltage value, synchronized switching harvesting on inductor technique is used. An electronic switch is synchronously commanded with the vibration and connects briefly the piezoelectric module to an inductance allowing voltage inversion and keeping voltage value stable.

The module performances have been evaluated by experience. A vibration harvester was mounted on a rod representative of the application. The assembly is put on a rig allowing to apply load and to heat the prototype to reach application temperature. An electromagnet is used to induce vibration in the piezoelectric beam. Vibration signal is a sinusoidal wave at 100 Hz, and several amplitudes are

DOI: http://dx.doi.org/10.5772/intechopen.86758

applied. Power is applied to a resistive load to measure energy produced by the harvester. **Figure 7** presents the harvester power characteristic. 4 mW is produced once a vibration level of 1 g is reached.

The requested power (4 mW) by the system is generated by 1 g acceleration level.

3.4 Thermal harvester

Thermoelectricity is based on the Seebeck effect: conversion of heat into electricity. This makes it possible to save heat that would otherwise be lost. A good thermoelectric generator (TEG) must have a high Seebeck coefficient to produce the required voltage, a high electrical conductivity to reduce thermal noise, and a low thermal conductivity to reduce thermal losses (see **Figure 8**).

The engine link can be used as the hot source temperature since it is connected to the engine which is hot in service. The air around it must be cooler than the strut load and will be used as the cold source temperature.

The module performances have been evaluated experimentally. The mechanical connection of the rod is linked to heating plate, allowing a thermal gradient equivalent to the application, expecting ambient temperature (**Figure 9**).

With the circuit load, the module provides for a gradient of 10°C 500 mW of power. Two modules are integrated to the system.

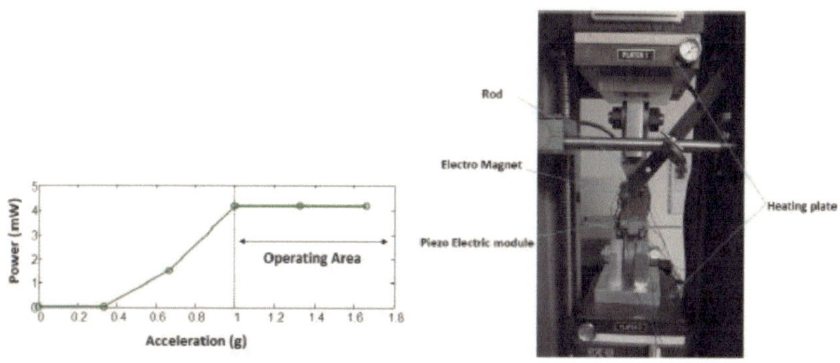

Figure 7.
Vibration harvester evaluation.

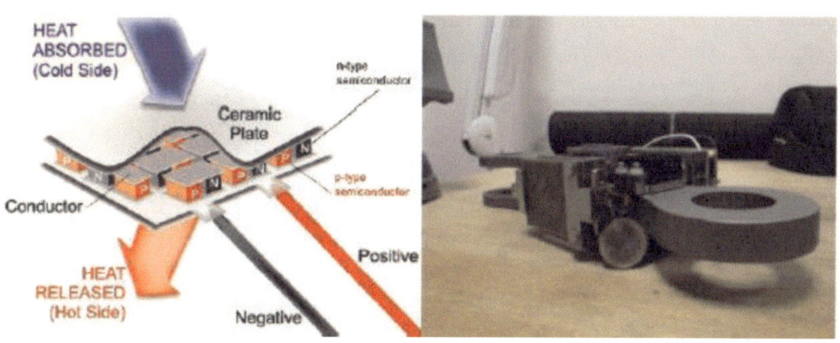

Figure 8.
Seebeck effect.

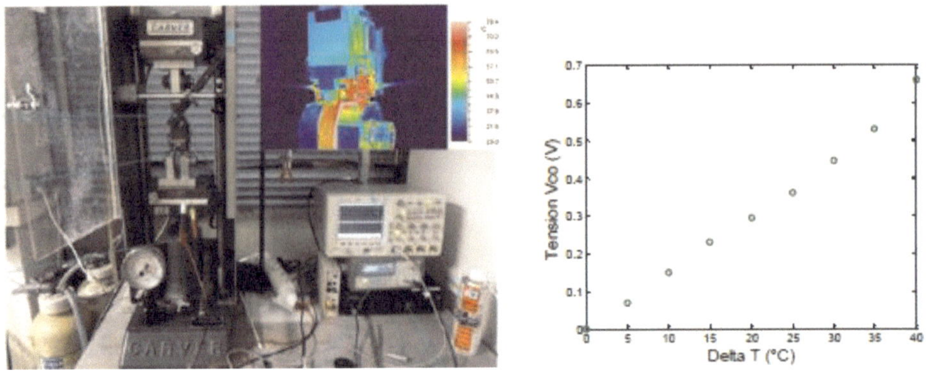

Figure 9.
Thermal harvester characterization.

3.5 Power conversion

Power management block is one of the main blocks of this module because it is supplying a whole circuit from available harvesting. It provides information to microcontroller (MCU) about the state of the tag. It indicates if vibrational, RF, or thermal energy is available. It also manages supply voltage for the whole system (**Figure 10**). Synchronized switch harvesting (SSHI) techniques are used to increase vibration harvester efficiency, as detailed in Ref. [5].

3.6 RFID as a power source

The internal impedance of the chip and the antenna is given by $Z_c = R_c + jX_c$ and $Z_a = R_a + jX_a$, respectively. Thus, the power transferred to the chip *P_chip* is related to the power in the tag *P_tag* by the following equation:

$$P_{tag} = P_{chip} * \tau$$

where τ is the power transmission coefficient:

$$\tau = 1 - \left|\Gamma_{tag}\right|^2$$

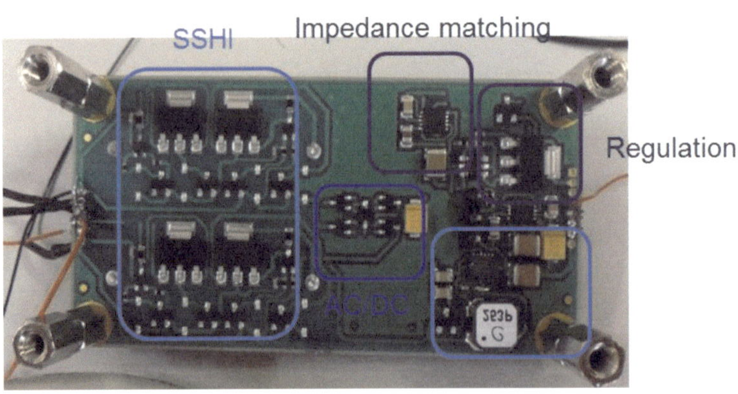

Figure 10.
Power conversion board.

And Γ_{tag} is the reflection coefficient from the antenna and the chip:

$$\Gamma_{tag} = \frac{\left(z_c - z_a^*\right)}{z_c + z_a}$$

So τ can be represented by the following equation:

$$\tau = \frac{4 * R_a * R_c}{|Z_c + Z_a|}$$

Increasing the value of τ will increase the power reaching the tag, increasing, consecutively, the distance from where the tag can be active by the amount of power coming from the reader.

Finally, τ is a coefficient that varies between 0 and 1. To increase the reading distance, the power transmission coefficient should be as much as possible equal to 1. That means that the impedances of the antenna and the chip should be matched:

$$z_c = z_a^* => \tau = 1$$

The theoretical distance from where the tag can be read is calculated using Eq. 1, where λ is the wavelength, P_{trans} is the power sent by the reader, G_{reader} and G_{tag} are, respectively, the gain from the reader and the tag antenna, and P_{th} is the tag chip sensibility, which is the minimal power that the IC must receive to be activated. The value of $P_{trans} * Greader$ is fixed as the maximum power that is possible to be sent by the local legislation in the desired frequency band.

$$Read\ distance\ r = \left(\frac{\lambda}{4}\pi\right) * \sqrt{\frac{P_{trans} * Greader * Gtag * \tau}{P_{th}}} \tag{1}$$

Commercial readers are normally very sensitive. The minimum power requested to decode without ambiguity the tag signal (the downlink) is considerably smaller than the one characterizing the uplink. Therefore, for passive UHF-RFID tags, the maximum distance from where the system can work is only governed by the uplink for most of RFID systems.

Far-field UHF-RFID antennas are usually designed in three steps: firstly a loop is designed around the IC to work as an inductor and compensate the capacitive part of the chip. The system composed by the loop and the chip will resonate around the desired UHF frequency. Secondly, the radiation element that is usually a folded dipole is attached to the antenna. Finally, the impedance matching and radiation gain for the tag are tuned and optimized to increase its reading distance. This step is very important in the way it permits a better understanding of the antenna electrical behavior. It is in that step that the designer can find a correspondence between the geometric dimensions and the EM properties of the antenna [6].

For some classical HF antennas, there are analytical formulas that can be used to calculate the expected EM properties of the coil in order to simplify the design part. However, for UHF-RFID antennas, these formulas are very uncommon, and most of the process of the antenna design is empirical. In [7] a method to automatically design UHF-RFID antennas by using a combination of a genetic algorithm and the 3D EM simulation software CST is given.

Besides the classical dipole antenna, there are several researchers that are trying to create new designs to surpass specific problems that most of the time accrued in practical applications. Nested slot, inverted-F configurations, and patch antennas are some of the most well-known designs [8].

Even if the UHF-RFID is being used in a large number of economic areas, there is already a large market where this technology did not bring any answer. Metal placed with RFID tags is one of the greatest challenges to the RFID technology, and a lot of researches have been putting large efforts to solve this problem in the last decade. In particular, the classical dipole antennas are not effective when put near to metallic surfaces.

This specific problem is explained in **Figure 11**, which shows that the image dipole below the metal surface has an opposite current from that of the original dipole. If the space between the dipole and its image is very small (much less than one wavelength), then the total effective current approaches zero. Therefore, the total radiation field is negligible. The RFID is then unable to capture power from the reader.

Looking for a solution to this problem, some ideas have been explored in the last years and are well explained in [6, 7]:

- Inserting high permeability isolator
- Increasing the distance between the antenna and the metal surface
- Using frequency-selective surfaces
- Using an anti-metal antenna design

Dipole antenna demonstrated weak communication capability for the project application environment. An anti-metal antenna design has been investigated [10–22] to propose an alternative to dipole antenna as shown (**Figure 12**).

4. Testing in a laboratory environment

Once evaluated in a laboratory environment, the demonstrator was integrated to engine tests. Packaging of the system was made of silicon sealant. A full-scale demonstrator was evaluated and submitted to qualification testing before ground engine testing. **Figure 13** presents the system validation phase:

- The demonstrator is assembled on a real thrust link.
- The thrust link is mounted on a hydraulic press able to generate compressive force similar to the real application.
- The press plates can be heated allowing to apply application temperature level to the system.

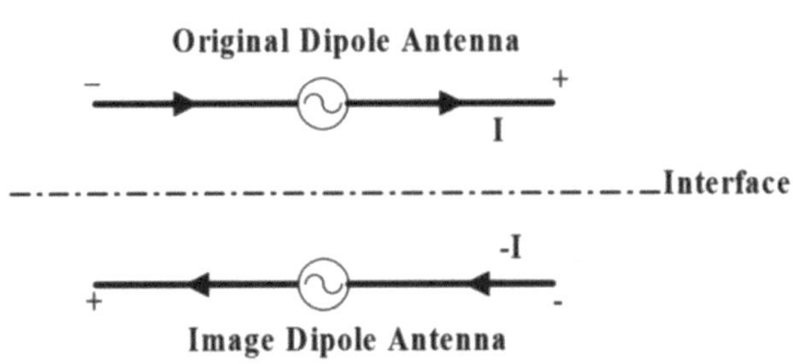

Figure 11.
A dipole antenna above a perfect electric conductor (PEC) and its image [6].

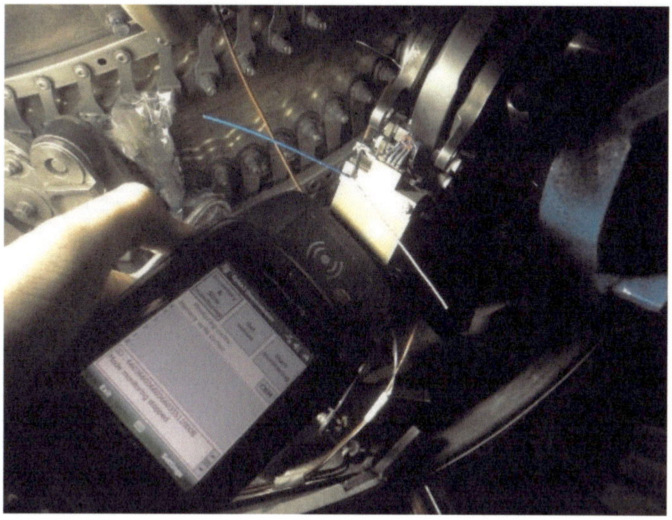

Figure 12.
Reading distance with dipole antenna.

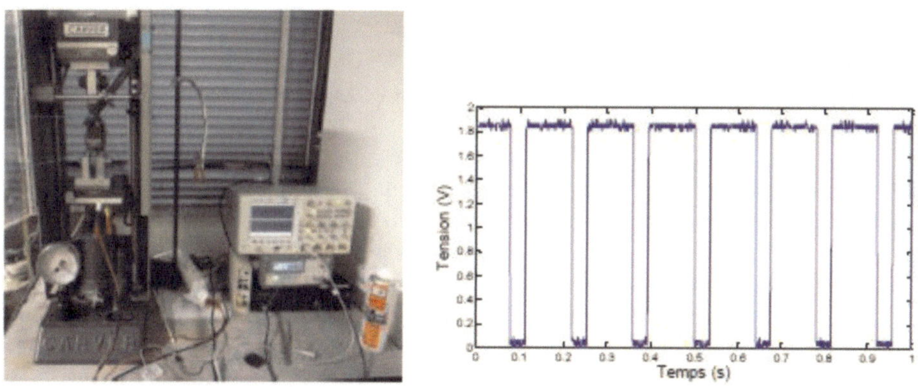

Figure 13.
Laboratory validation of the sensor system.

- A vibrator is used to induce vibration to the piezoelectric beams of the vibration harvesters. Levels applied are random spectrum, according to application specification.

Compressive loads were applied to the mechanical link with the system clamped to it. Thermal difference (80/120°C) was applied at each end of the link to establish a thermal gradient close to the application to feed thermal harvesters. A piezoelectric buzzer was used to apply excitation to the vibration harvesters. The regulated voltage from the power module was checked and also the read write memory voltage output at the microcontroller level.

The autonomy of the system was reached for a thermal gradient of 10°C for the thermal harvester and a vibration level of 1 g at 100 Hz (**Figure 14**).

The system was submitted to ground testing inside a real engine. During the tests, system failed to be autonomous. Measurement done externally showed that the vibration and energy sources were lower than forecasted for the sizing of the system.

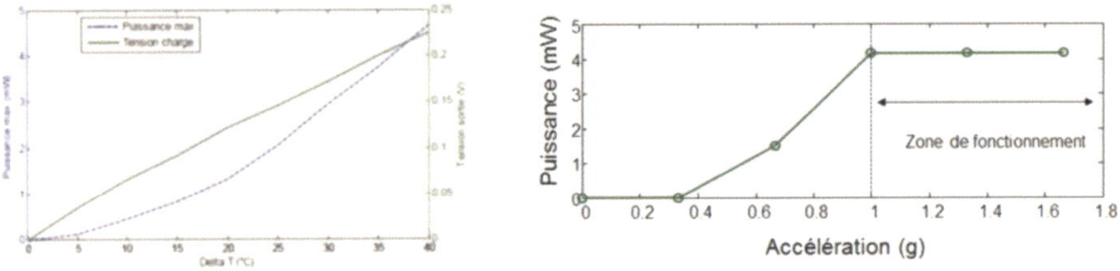

Figure 14.
Thermal and vibration harvesters' characteristics.

5. Conclusion

Optimization of energy harvesters is a system-level problem that involves several design requirements on the power processing stages. Deploying an energy harvester on its own will yield poor power densities, which is why additional circuitry is needed to implement features such as synchronized switch harvesting or impedance match between harvester and load electronics, energy storage capabilities, and output voltage regulation.

Each energy harvester is differentiated by its transduction mechanism, and therefore the equivalent source impedance model must be derived for different harvesters and available energy densities variation from the environment. By matching the source impedance to that of the load or by applying appropriate switching, the maximum power transfer is achieved from the harvester to the load under optimal conditions **Figure 15**.

The project provided key data to identify the gaps to be solved for developing energy harvesting module in engine environment.

Despite several energy sources available in airplanes and helicopters, engineers and scientists have to face numerous challenges when trying to achieve reliable devices able to capture sufficient energy to perform any useful work.

Energy sources available must be characterized in terms of level, availability, and frequency range. This is a prerequisite before sizing the harvester module to ensure capability to produce enough energy.

Energy management is also key: the best practice approach is to use a comparator on the energy storage module to prevent the system from drawing energy from the storage element, unless enough energy was stored to perform the programmed sensing/processing/logging communication tasks. This allowed the energy storage elements to store up enough energy to perform for the specific task, before allowing

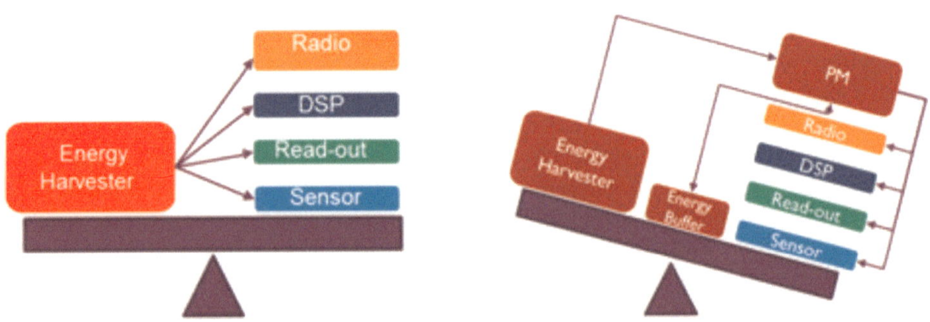

Figure 15.
Constraints in designing an autonomous system.

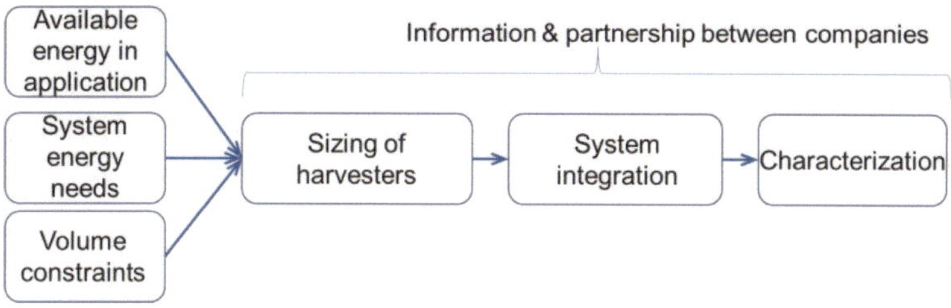

Figure 16.
Ideal workflow for developing an autonomous system.

the task to be performed. This is the key approach to adapt the sensing tasks in applications where the ambient energy levels are low, variable, or intermittent.

Energy harvesting based on vibrational and thermal effects has been chosen for the module development, thanks to their better reliability and performances than others. In the case of vibrational energy, the production of micro-generators using either electromagnetic or piezoelectric conversion elements seems the most promising. Both conversion principles are complementary although the electromagnetic one is more effective in the low frequency range (100 Hz).

For thermal energy harvesting, the system operation is based on the principle of the Seebeck effect. Such technologies are investigated because of their robustness and resistance to environmental stresses, i.e., mechanical and thermal. Nonetheless, a number of challenges must be solved before achieving efficient wireless autonomous sensors for real aeronautical applications, such as long-lasting self-sufficient operation of sensors, selection of wireless protocols, etc.

According to **Figure 16**, the goal of maximizing the amount of the harvested energy involves several factors, including electronics optimization, characterization of the available ambient energy, selection and configuration of energy harvesting materials, and integration with storage mechanisms, along with the power optimization and power awareness design. This project tried to address these issues in an integrated manner from the multidisciplinary engineering perspective. The performance of thermal and vibration harvester prototypes had been validated and tested in real environment.

Future perspectives are optimization of the transduction mechanism of the harvester, power awareness, and storage element to validate an upgrade of the prototype in experimental environment derived from the real measurements.

Acknowledgements

Work in the project has been done in the EPICE-CORALIE program, a funded project from French CORAC. Project was done in partnership with SAFRAN.

Experimental work was supported by LGEF-INSA Lyon, Université de Lyon.

Author details

Yoann Hebrard
SKF-Aerospace, Valence, France

*Address all correspondence to: yoann.hebrard@skf.com

References

[1] Lacaille J. Standardized failure signature for a turbofan engine. In: IEEE Aerospace Conference; 2009

[2] Bense W. Prognosis and health monitoring systems for aircraft engines. In: SAE Aerotech; 2013

[3] http://www.boeing.com/commercial/aeromagazine/aero_14/conditional.pdf

[4] Le MQ, Capsal J-F, Lallart M, Hebrard Y, Van Der Ham A, Reffe N, et al. Review on energy harvesting for structural health monitoring in aeronautical applications. Progress in Aerospace Sciences. 2015;**79**:147-157

[5] Lallart MN. Amélioration de la conversion électroactive de matériaux piézoélectriques et pyroélectriques pour le contrôle vibratoire et la récupération d'énergie: Application au contrôle de santé structurale auto-alimenté. In: INSA Lyon; 2008

[6] Lallart M. Small-scale energy harvesting. 2012

[7] Sebald G, Guyomar D, Agbossou A. On thermoelectric and pyroelectric energy harvesting. Smart Materials and Structures. 2009;**18**(12):125006

[8] http://www.rfidc.com/docs/rfid.htm

[9] http://www.centrenational-rfid.com/les-gammes-de-frequences-rfid-article-16-fr-ruid-17.html

[10] Li Y, Rongwei Z, Staiculescu D, Wong CP, Tentzeris MM. A novel conformal RFID-enabled module utilizing inkjet-printed antennas and carbon nanotubes for gas-detection applications. IEEE Antennas and Wireless Propagation Letters. 2009;**8**: 653-656

[11] Nair R, Barahona M, Betancourt D, Schmidt G, Bellmann M, Höft D, Plettemeier D, Hübler A, Ellinger F. A fully printed passive chipless RFID tag for low-cost mass production. In: The 8th European Conference Antennas Propagation (EuCAP 2014); 2014. pp. 2950-2954

[12] Vena A, Perret E, Tedjini S, Tourtollet GEP, Delattre A, Garet F, et al. Design of chipless RFID tags printed on paper by flexography. IEEE Transactions on Antennas and Propagation. 2013;**61**:5868-5877

[13] Finkenzeller K. RFID Handbook. 3rd ed. Chichester: John Wiley & Sons, Ltd; 2010

[14] Mandeep K, Manjeet S, Neeraj M. RFID technology principles, advantages limitations & its applications. International Journal of Computer and Electrical Engineering. 2011;**3**(1):166-178

[15] Zhan S. Analysis and design of metal-surface mounted radio frequency identification (RFID) transponders. In: Iowa State University; 2008

[16] Björninen T, Sydänheimo L, Ukkonen L, Rahmat-Samii Y. Advances in antenna designs for UHF RFID tags mountable on conductive items. IEEE Antennas and Propagation Magazine. 2014;**56**

[17] Zhan S, Weber R. A low profile metal-surface mounted RFID tag antenna with a slotted inverted-L shape. 2007;**10**:213-218

[18] Zhan S, Weber R. An ultra-thin and high gain metal-surface mounted RFID tag antenna with a symmetrically slotted patch

[19] Zhang J, Yunliang L. A novel metal-mountable electrically small antenna for RFID tag applications with practical guidelines for the antenna design. IEEE

Transactions on Antennas and Propagation. 2014;**62**:156-168

[20] Elboushi A, Haraz O, Jamil K, Sebak A-R. Analysis and design of a metal-backed RFID tag antenna. In: IEEE International Symposium on Antennas and Propagation & USNC/URSI National Radio Science Meeting; 2015

[21] Petrariu A-I, Popa V, Chirap A. Metal mountable microstrip patch UHF RFID tag antenna. In: International Symposium on Signals, Circuits and Systems (ISSCS); 2015

[22] Faudzi NM, Rashid ANA, Ibrahim A, Khyasudeen MF, Ali MT. Microstrip Dipole UHF-RFID

Chapter 3

The Importance of Emissivity on Monitoring and Conservation of Wooden Structures Using Infrared Thermography

João Crisóstomo and Rui Pitarma

Abstract

Much of the built heritage is built of wooden structures. Due to the lack of maintenance, it is susceptible to biological attacks, such as fungi and wood destroying insects. Most of the methods used for its inspection and evaluation are intrusive. More friendly methods are required. Infrared thermography, being a non-destructive, contactless and versatile technique, can be a very useful tool in this field. However, the correct temperature measurement depends greatly on the emissivity value of the material. In this chapter, the emissivity values are presented and discussed for wood samples of *Pinus pinaster* species. In a qualitative analysis, this factor is not so important. Moreover, in a quantitative analysis for which the measured temperature value is relevant, it is crucial to know the emissivity value.

Keywords: emissivity, infrared thermography (IRT), wood, assessment of wooden heritage structures, heritage conservation, non-destructive testing, quantitative analysis, sustainability

1. Introduction

Throughout the world, wood is a universal material used since antiquity in infrastructures and structures. Like other building materials, wood develops pathologies that affect its durability and useful life expectancy. Humidity and temperature variations of indoor and outdoor environments are main factors that affect the biological degradation rate of wood. Wood deteriorates more rapidly in heat and humid than in cold and dry environments. Main organisms that degrade wood are fungi and insects, causing damage ranging from simple discolouration to complete deterioration; the damage can be possibly aesthetic or it can have disastrous consequences, such as the collapse of the structures. To prevent and control damage, it is crucial to evaluate the state of conservation of building elements. It includes the necessary non-destructive evaluation of its mechanical properties and conditions of wood [1]. Wooden structure monitoring is not necessarily expensive and prevents loses.

The early detection of anomalies allows avoiding aggravated damages in the structure that can lead to the inevitable replacement of structural elements. It is, therefore, relevant to prevent harm to people and property. It is essential that inspections of structures be carried out regularly in order to monitor the state of conservation.

The need to solve practical problems without destroying the integrity of the object under inspection has motivated the development of measurement techniques for assessing the physical properties of lumber. The earliest non-destructive assessment is a visual inspection; it is also widely used for wood product classification [2]. In general, the techniques currently used provide information at specific points. The whole is obtained by extrapolation, with a long series of studies. Often, the structure to be examined cannot be reached from the ground. They are time-consuming diagnostic methods that require the presence of teams of technicians. They are usually invasive because they require holes to be made which may also become access routes for pathogens. X-ray methods are also used and are therefore potentially harmful to health. The most frequent method with the best cost/benefit is the percussion of the wood using a blunt object; the analysis is based on the interpretation of the sound produced. The results, however, depend on the operator's experience and require direct contact with the area to be inspected, so it is necessary to mount scaffolds or staging when the elements under analysis are not reached from the ground. The interpretation of the information can be slow and difficult, depending on the variety and quantity of elements. This kind of inspection makes it difficult to record data to be compared to subsequent evaluations.

Thermography is a generic term for a variety of techniques used to visualise the temperature of object surface. Thermal imaging is the result of a complex interaction between the heating source, the material and its irregularities. In fact, the thermal properties of the materials are conditioned by the degree of their structural integrity; that is, when there are anomalies in the material structure, the heat flux changes. Then, thermography detects these changes. The surface thermal mapping results in the thermogram. The thermogram indicates the location where thermal heterogeneities exist. Therefore, it shows if a structure has damaged or defective parts. The thermogram displays the temperature variations using colour gradient.

Thermography is an *in situ* technique based on thermal image examination that can be applied to wood structures to find external signs which might indicate possible internal deterioration. This method is non-invasive and totally harmless to people. Being a non-destructive and non-contact technique, it can be a versatile tool, very useful for inspection. A portable infrared camera can be used to evaluate the structure in real time even when the structure cannot be accessed from the ground. Another advantage is that this technique allows the monitoring of disease progression. It makes possible to detect water content, state of deterioration, loss of density and anomalies. The assessment of the wood condition gives the information for maintenance and repairing.

In the science of wood, infrared thermography (IRT) is a relatively recent field of study [3]. The two main procedures are passive and active thermography. It is suitable for wood diagnose because it is a non-destructive and non-contact technique for recording the temperature of object surface, based on emitted radiation. The temperature values are mapped using false colour patterns.

In most of the instruments that use this technology, it is necessary to introduce an emissivity factor to calibrate the temperature measurement. The emissivity together with the reflected temperature allows the acquiring of data to produce the thermographic image [4]. The main issue is the determination of specific wood emissivity values.

1.1 Active and passive procedures

Maldague describes two thermographic procedures. In the active thermography, an energy stimulus is artificially produced on the object of study to cause an internal flow of heat on the surface to be inspected [5]. This flow can be triggered by different processes: simple thermal sources like lamps, heaters or flashes, hot air jets, ultrasonic pulses, infrared radiation, microwave, laser and among others. The generated heat flux is disturbed if there are defects or damages on or near the surface of the object of study. It causes discontinuities and thermal contrast. They are detected by analysing the thermograms obtained during the thermographic inspection. In the passive procedure, the thermal contrast is generated by natural sources such as sunlight [5–7].

1.2 Qualitative approach versus quantitative approach

The thermographic analysis can be qualitative and quantitative. Qualitative thermography focuses on the study of thermal patterns to reveal the existence and location of anomalies. Quantitative thermography uses temperature measurement as a criterion to determine the severity of an anomaly and to set repair priorities. Applying the qualitative approach, it is possible to determine the existence and location of a problem in the wood. The quantitative approach determines the severity of the problem; it can still help to determine when it should be repaired, by quantifying the temperature [6]. However, an accurate temperature measurement is strongly dependent on the value considered as a reference for the emissivity of the material.

The influence on temperature by the variation of the emissivity value can be observed in **Figure 1**. This figure shows the temperature reading in the thermographic camera as a function of the emissivity introduced in the equipment. When the calibration value of 0.89 (emissivity of the surface of the wood measured with the black tape method) is entered, the average value of the surface temperature is 43°C. However, if a different emissivity value is entered in the thermographic camera, the value of the temperature reading will deviate significantly from the actual value of the surface temperature. This error is all the greater the error in the emissivity value. For example, if an emissivity of 0.40 is introduced, we obtain a mean temperature value of 61°C, significantly different from the actual value of 43°C.

1.3 Factors affecting thermographic inspection

IRT is a measurement technique based on the detection of radiation in the infrared spectrum. In fact, all bodies above 0 Kelvin (absolute zero) emit this electromagnetic radiation. Electromagnetic radiation detection in the infrared (IR)

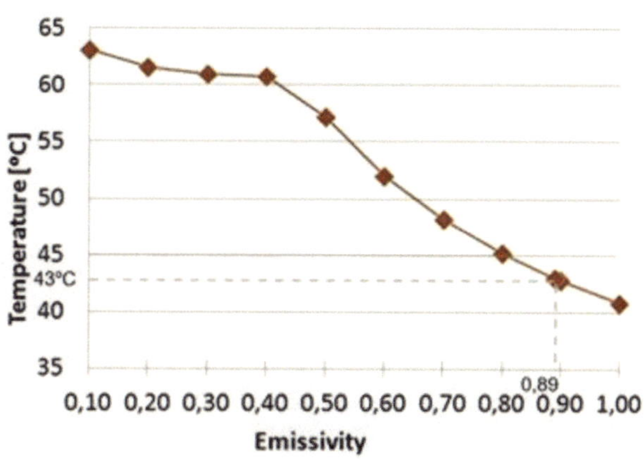

Figure 1.
Surface temperature of the wood sample as a function of the emissivity value.

spectrum is usually done between 2–5.6 μm and 8–14 μm. Both spectral bands are generally used due to their atmospheric malabsorption [5].

There are building materials with high emissivity in both spectral windows, as white marble. But there are other materials, such as wood, which have a wide range of emissivity due to the spectral window used in the observation [4]; therefore, it should be mentioned in each observation. Rice further states that the shorter wavelength spectral window (close to 0.7 μm) is desirable because the effective emissivity is higher and the effect of the surface characteristics is minimised [8]. Dewitt and Nutt (1988) in Rice highlight that the angle of observation in IRT is also determinant [8]. Radiation from a surface occurs in all directions, so the hemispherical term is often added. Mathematically, the solid angle on which the radiation measured is often referred to. However, in real surfaces, the radiation is not uniformly reflected in all directions. Then, the preferred method is to measure the normal or near normal radiation of the object surface. Fronapfel and Stolz conducted studies to determine the emissivity with different angles of observation where the emissivity variation can be observed [9]. This technique for building monitoring is expanded and has been used for more than 25 years in the diagnosis of any kind of buildings: historical, monuments and modern structures. Passive thermography is a qualitative method and is commonly used in the investigation of buildings. The purpose is to detect irregularities and malfunction. However, the lack of specific emissivity values of each material makes accurate temperature measuring impossible. In fact, there is not an infrared camera that reads directly dispensing this factor. All cameras interpret infrared radiation from the surface taking into account emissivity, reflectivity and, occasionally, the transmissivity of IR radiation. The transmissivity is only important for large distances between the camera lens and the object of study [10].

Figure 2 shows examples of the influence of homogeneous heating and adequate focusing in the determination of an accurate emissivity value [7]. Even with the correct emissivity value, incorrect focus or improper brightness and contrast may cause the observed surface temperature readings to be incorrect. For an accurate determination of the surface emissivity, it is necessary that the temperature of the surface is to be between 10 and 20°C above the ambient temperature. It is also important homogeneity in the heating process. **Figure 2** shows cases of thermograms where the heating was neither homogeneous, nor a correct adjustment in the focus, which led to the determination of incorrect values of emissivity and surface temperature.

1.4 Emissivity

Emissivity plays an essential role in thermography. Emissivity is used to characterise the optical properties of materials taking into account the amount of energy emitted compared to an ideal blackbody [10].

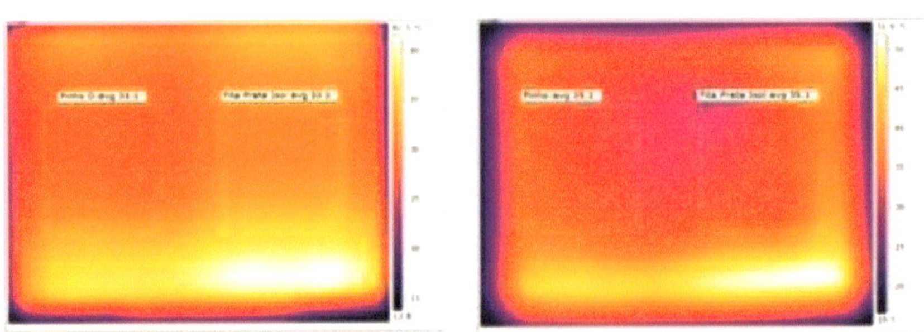

Figure 2.
Thermograms that show the heterogeneity of surface heating and inadequate focusing.

The emissivity scale goes from 0 (perfect mirror reflector) to 1 (perfect emitter black body). It depends on temperature, wavelength and surface conditions, such as roughness. Some authors also refer to the temperature reflected as being mainly dependent on the radiation coming from the surrounding environment as another factor that affects the accuracy of the temperature measured with infrared thermography [10, 11]. A surface with a low emissivity value, such as aluminium, steel, etc., acts as a mirror (high reflectance). However, these problems are usually solved using tapes or high emissivity paints [10].

Most of the materials used in buildings have high emissivity, generally above 0.8 [10].

IRT applied in building diagnostics can be used efficiently in the detection of heat loss, quality of thermal insulation in walls and roofs, thermal bridges, airflows and sources of moisture [6, 10, 12]. It also mentioned the capability of identifying structures, holes, cracks, material discrimination, analysing the state of mural paintings and the state of conservation of materials [13].

Avdelidis and Moropoulou determined emissivity values in historic buildings at different temperatures, applying infrared thermography of medium and long wavelength [10].

1.5 Thermography applied to wood structures

The literature describes the use of IR thermography for the evaluation of wood structures in historical buildings, such as the San Felipe Neri Oratory (Cadiz) [14] and a mosque in Ankara, Turkey [15].

Rosina and Robison describe that in Italy, IRT was used in different phases of building rehabilitation, from preliminary investigation to the rehabilitation process itself, as a final inspection instrument or as a part of the periodic observation and maintenance plan. This technique allows researchers to gather information about the location, shape, characteristics of the materials, and deterioration state of the elements and constructive systems. Taking into account the particular boundary conditions, the distribution of the surface temperature of the object of study allows the detection of discontinuities and changes in the structures of the buildings [6].

Studies on wood have confirmed the role of certain factors that affect its surface temperatures such as density, nodes, biological deterioration and cracks [16]. As referred above, the emissivity plays a crucial role in the thermography analysis. Lopez et al. determined the emissivity of wood samples using different species that were exposed to different temperature values. They propose an average emissivity value of 0.924 for temperature 22°C and spectral band 7.5–13 μm. They suggested that this value can be used for any type of wood, minimising the possibility of error in the determination of surface temperatures [17].

For an IR thermography test, it is necessary to have a thermal gradient to induce the thermal response of the surfaces. It is considered that 20°C is a satisfactory difference [18], while ASTM E1933-99a refers to a minimum of 10°C [19]. The ASTM E1933-99a standard is an existing normative that establishes procedures for the measurement and determination of emissivity of various materials using IR imaging systems [19].

Despite the theoretical and practical knowledge about this technique in several application areas, it has only recently been used for the diagnosis and evaluation of wood structures.

Thermography is used because the thermal conductivity and the resistivity of the materials are related to the degree of their integrity. That is, the structural anomalies cause a change in heat flow. Thermography indicates if a structure has damaged or defective regions; it shows the location of thermal heterogeneities. The thermograms display the different local temperatures using colour gradients; they are monochromatic scales, such as greyscale used at the beginning or the current polychromatic scales. In the thermograms of **Figures 3–5**, thermal heterogeneities can be observed denoting material cracks, both in the active and passive modes. Generally, in the passive mode, the cracks are noticed by lower temperatures, whereas in the active mode they are evidenced by higher temperatures [7].

Thus, **Figure 3** shows a thermogram obtained in the passive mode. It was taken from a wooden structure of a roof. A crack may be seen running longitudinally through the wooden beam. The crack is visible because thermal heterogeneity is visible in this part, that is, temperature values lower than the overall beam pattern, visible through darker shades. This information is obtained by comparison with the colour/temperature scale that is displayed on the right side of the thermogram.

Figure 4 shows a thermogram of the face of a sample of wood previously heated in an oven (active mode). This figure shows that in the active mode, cracks are evident because they exhibit thermal heterogeneities with temperature values higher than the general pattern of the observed object face. In this case, it is noticed by the lighter colours, unlike what happened in the passive mode of **Figure 3**.

Figure 5 shows the conventional photograph (visible wavelength) of a wood surface and its thermogram (infrared wavelength) in active thermographic mode. By the active procedure, this type of nodes and the cracks show colour heterogeneity in relation to the general pattern denoting higher temperature. The cracks of **Figure 5** although they are barely perceptible in the photograph are well seen in the thermogram.

Thermography also plays an important role in the inspection, assessment and monitoring of living wood structures, we mean trees. There are living wooden structures with relevant historical and cultural importance as the monumental trees and the notable ones. Thus, there are studies carried out about how to inspect, analyse and evaluate them to be safeguarded. Research on living wood structures has already been carried out in Italy. Since the 1980s, Giorgio Catena and Alexandra Catena have validated the usefulness of IR thermography technique [20]. In

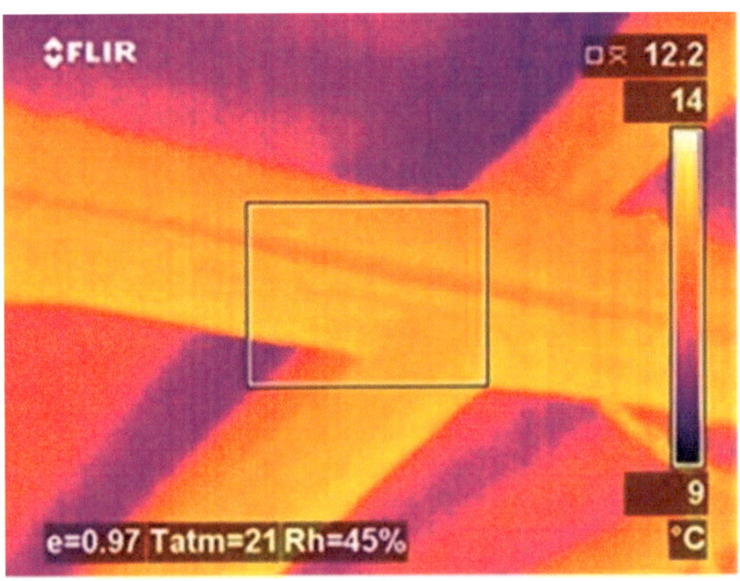

Figure 3.
Thermogram of old wooden structure: the crack is visible along the beam (passive method).

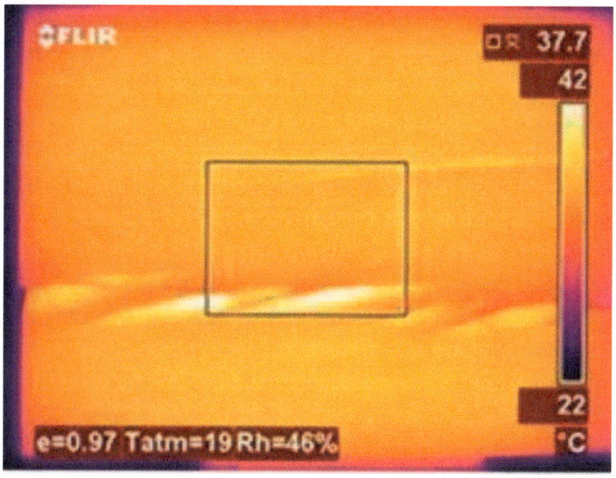

Figure 4.
Thermogram of a wood sample with visible cracks (active method).

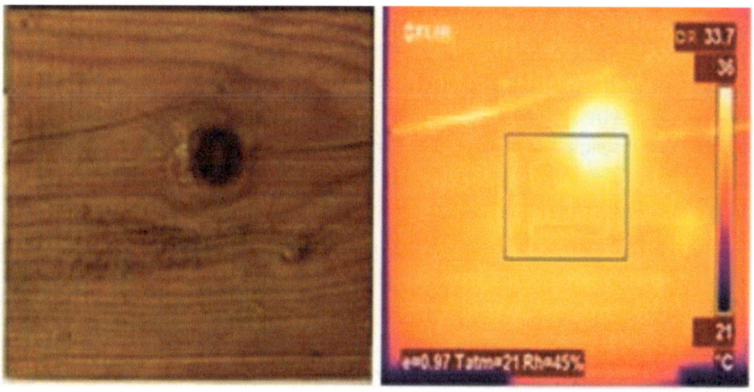

Figure 5.
Photograph and thermogram of pine sample (active mode). It shows cracks and knot.

England, Marcus Bellett-Travers applied the IR thermography to inspect, evaluate and minimise wood structure risk [21].

In Portugal, the TreeM Project aims to disseminate the possible uses of this technique [22, 23].

It is essential to safeguard, for as long as possible, the lives of these incredible living beings. Some trees are catalogued for conservation and protected by law due to their age, shape and size. Periodic inspections are also required in places of high risk of partial or total collapse. This is a mean to prevent losses to individuals and property.

2. IRT monitoring for the conservation of wood structures

IRT is a method of inspecting structures, whose data (thermograms and radiometric videos) require visual inspection with photographs in the visible range. Visual inspection is critical as certain heterogeneities detected by IRT may be non-significant, resulting only from, for example, dust. It should be noted that many of the present-day thermographic cameras already include a photographic camera in order to carry out the process in the same procedure, thus providing a photograph and a thermogram of the object with the same perspective.

The analysis of the information obtained provides elements for an evaluation of the structure, that is if the structure of wood presents or not anomalies (e.g., due to excess moisture, by the existence of fungi or by wood-decaying insects). This inspection shall be periodically repeated over the life of the structure so that the faults detected in the meantime are corrected in a timely manner. Thus, more radical and costly measures are avoided, extending the useful life of the structure.

The comparison of thermograms obtained over time is possible if a suitable value of emissivity of the surface of the wood studied is used. Some examples of the application of IRT in the detection of biodegradation are given by Grossman [24, 25]. A qualitative inspection requires the entering in the thermographic camera of a high emissivity value, since the wood is a material of low reflexivity and therefore with high emissivity. When inspecting the wood structure, non-conformities can be observed in the thermal pattern; they are identified by differences in the colour pattern in some places in relation to the whole.

In the case of quantitative analysis, it is relevant to know the concrete value of the temperature in order to have an idea of the severity of the damage detected. Therefore, a correct emissivity value is required, which is dependent on the environmental conditions of the place where the observation is being carried out. It is difficult to determine the correct emissivity value for an *in situ* given surface. This difficulty results from two fundamental aspects: (a) the difficulty in reaching the structure and (b) the difficulty in raising the temperature of the wood of at least 10°C [19] at 20°C [18] above ambient temperature because the higher the temperature differential, the greater the accuracy of the determined emissivity [19]. Thus, it is crucial to have tables of emissivity values for different types of wood to select the value that approximates the most to the real conditions. It would avoid the need of determining this value each time a study is carried out. Kang et al. emphasise the importance of tabulated emissivity values proposing a study with hundred species from Korea and Tropics [26].

The value of the emissivity is also a function of the density of the wood. If we use the same emissivity value, different temperatures will be seen in the wood. The deterioration causes a decrease in density. In the thermogram, this anomaly is detected through thermal heterogeneity indicating the points of deterioration.

Main advantage of IRT is that it indicates locations that require more detailed inspection and, if necessary, the application of other inspection techniques. The IRT does not replace the use of other techniques for evaluating structures, but provides information to apply other techniques only in concrete and localised points, avoiding their application in unnecessary areas. IRT inspections can be performed from the ground observing the structure as a whole. Thermograms and radiometric videos made with latest thermographic cameras produce records that can be compared with records of previous and future inspections.

3. Experimental work

3.1 Objectives of experimental work

The objective of the present study was to determine the emissivity values of two different samples of sawn wood with similar textures and colouration of the species *Pinus pinaster* Aiton (commonly known as brave pine or maritime pine). Although it is of same species, the two samples have different densities, since they come from two different locations with different climates—one from the region of Leiria (coast) and the other from the Serra da Estrela region (inland mountains). It is therefore intended to contribute to the need of tables of emissivity values for different wood species, in environmental conditions close to those *in situ* observed.

3.2 Instruments and materials

A Velleman DVM401 Digital Multifunction thermohygrometer was used to determine the environmental conditions. The thermographic machine used was the FLIR ThermaCAM B20 with 36 mm lens, field of view (FOV) of 22.6°, whose first three digits of the serial number are 234. The minimum focus distance is 0.3 m. For a length of 0.5 m between the machine lens and the sample surface, it has a 0.2 m horizontal field of view (HFOV), 0.15 m vertical field of view (VFOV) and 0.63 mm instantaneous field of view (IFOV). The thermographic camera works in the spectral band of 7.5–13 μm, with a thermal sensitivity of 0.10°C at 30°C and accuracy of ±2°C and ±2%. The focusing is performed manually. The detector type is an uncooled microbolometer focal plane array (FPA) with a resolution of 320 × 240 pixels. ThermaCAM QuickView 1.3 [27] and ThermaCAM Reporter 7.0 [28] were the software used. The determination of the emissivity was done applying the black tape method; it used black adhesive tape ISO Tape Tesa. A WTC binder oven was used to increase the temperature of the samples. The electronic balance used for weighing the samples has a 5 kg maximum capacity and accuracy of ±1 g. The fixation of the sample and the thermal camera was done using two tripods.

3.3 Samples

The wood samples were of the species *Pinus pinaster* Ait, commonly known as maritime pine or brave pine. This species is common in Portuguese territory. The two samples came from different locations, with different climates and consequently different densities. This species of wood was commonly used in Portugal. Currently, it is found in most part of the built heritage. Roofs, floors and walls components were built with the use of brave pine wood. Since the end of the 1990s, glued laminated products were produced in Portugal and with them other species were commercialised such as spruce tree. Pine became rarely used for construction.

The lumber samples are unfinished wood, as is often found in building structures. The dimensions of the samples are: 20 cm long along the fibres, 15 cm in the transverse dimension and 15 cm in thickness. These dimensions were defined to simulate a small section of a current beam and to minimise the effect of the environment on the surface of the object of study. **Table 4** shows the sample condition as density and water content. **Figure 6** shows the images of the two specimens.

3.4 Experimental methodology

Two studies were carried out, one for each sample. In each study, two trials were performed. A thermal differential of approximately 20°C was adopted, as suggested in the FLIR ThermaCAM B20 manual [18]. The experimental work was carried out in one laboratory of the Polytechnic Institute of Guarda (IPG). The values determined for the emissivity were compared to those found in the literature.

Since it used a thermal imaging camera and software from FLIR manufacturer, the methodology suggested in the Camera manual [18] also mentioned by Spencer et al. [29] and ASTM E1933-99a (Reapproved 2010) was applied [19]. Some adaptations for the determination of the Reflected Apparent Temperature (Reflective Method) and for the determination of the emissivity were introduced.

For each of the two samples, it was observed the face parallel to the wood fibres (20 × 15 cm) with fewer irregularities, avoiding cracks and knots. The determination of the emissivity was carried out according to the methodology of the black insulated electrical tape (i.e. a material with known emissivity value). Strips of black adhesive tape (with an emissivity of 0.970) [18, 27] were placed juxtaposed

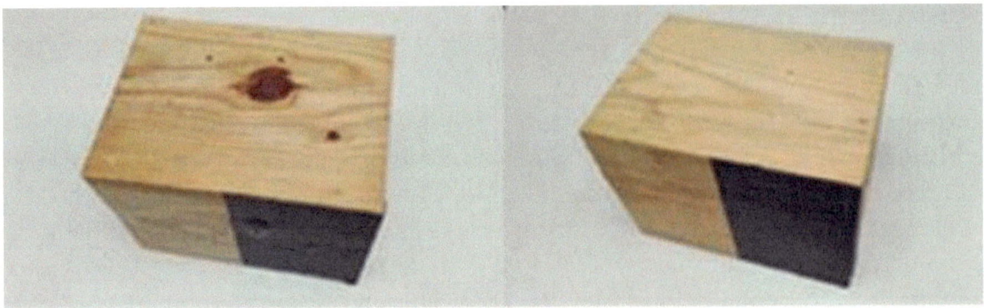

Figure 6.
Samples of Pinus pinaster Aiton: from Leiria (coast)—left side and from Serra da Estrela (inland mountains)—right side.

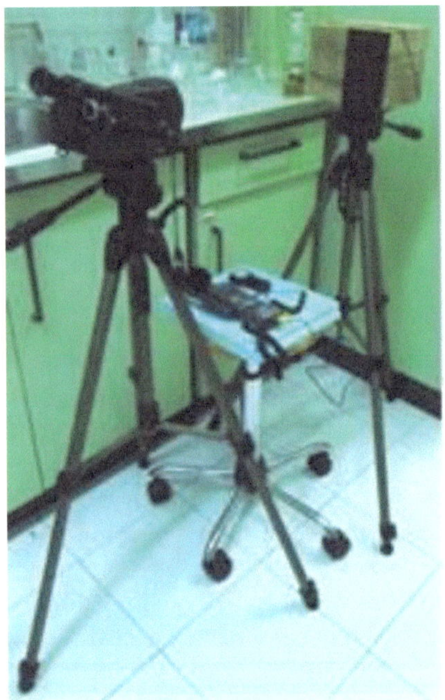

Figure 7.
Laboratory experiments (general view).

in the direction perpendicular to the wood fibres so as to fill half the face of the sample, that is, 10 × 15 cm.

The samples were heated at least 20°C above the temperature of the space where the observations were made. Thus, to ensure thermal uniformity, the samples were placed in the oven at 60°C for 24 h. A type K thermocouple made it possible to verify the same temperature in the surface of the two zones, the area of the black ribbon and the area of exposed wood.

In **Figure 7**, we can observe how the thermograms of the wood samples were made.

The tests were performed with low light intensity (see **Table 1**) to avoid reflections. It was found that for a low light intensity, the reflected temperature was near the ambient temperature. The distance between the camera and the sample was 0.5 m. A tripod was used to fix the sample so that it was observed perpendicular to the plane of the surface. The laboratory conditions relevant to the machine calibration were monitored: ambient temperature, relative humidity and light intensity, as shown in **Table 1**.

Place	Room temperature (°C)	Relative humidity (%)	Reflected temperature (°C)	Light intensity (Lux)
IPG	21.0	45	21.5	2.85

Table 1.
Laboratory conditions.

Measures were taken at a sole point lead to unrepresentative temperature readings and should be avoided. Since wood is a very heterogeneous material, thermographic evaluations should be applied to a relatively large area of the sample.

The thermograms were analysed with the aid of ThermaCAM Reporter 7.0 software [28]. It was determined the emissivity value of the sample exposed surface. The value obtained corresponds to the emissivity of the analysed wood sample at ambient temperature. This procedure was repeated two times for each sample.

Each sample was weighed to determine the water content [Eq. (1)]. Each sample was weighed after the heating process, wet weight (W_{Wet}). The determination of the water content of the samples followed NP 614: 1973 [30]. The samples were heated at 100°C for 48 h. Successive weight measurements were performed until the weight remained constant, dry weight (W_{Dry}).

$$\text{Water content [\%]} = \left((W_{Wet} - W_{Dry})/W_{Dry}\right) \times 100 \quad (1)$$

4. Results and discussion

The thermogram in **Figure 8** is an example of the methodology followed to determine the sample emissivity.

Table 1 shows the environmental conditions of the laboratory while the thermograms were produced.

Tables 2 and **3** show the emissivity and the surface temperatures for each thermogram of both pine samples. The samples are from Leiria (coast) and Serra da Estrela (inland mountains).

Table 4 shows the average emissivity of both samples of wood of the species *Pinus pinaster* Ait.

The emissivity of each sample results from the average of the two tests performed in each. They are 0.864 for Leiria pine and 0.873 for Serra da Estrela pine. In each case, the emissivity and temperature values were determined by the average area of one square on the software used (100 × 100 pixels). The results are valid for the environmental conditions and the sample conditions described in the previous tables.

It should be highlighted that there is no significant range difference between the two samples: 0.018 for Leiria pine and 0.035 for Serra da Estrela pine. On the other hand, the emissivity values are in accordance with the bibliography consulted, that is, the highest values are around 0.9: 0.855–0.873 for Leiria pine; 0.855–0.890 for Serra da Estrela pine. As reported by Rice [8], manufacturers of IR temperature measuring equipment often recommend values between 0.94 and 0.95 for generic wood and conditions.

The manufacturer FLIR indicates for four samples of pine in the spectral window 8–14 μm values of emissivity, namely between 0.81 and 0.89. The spectral window 2–5 μm indicates values of emissivity between 0.67 and 0.75 [18]. Holst refers 0.85 for the emissivity of wood planed in the spectral window of 8–14 μm [31]. However, it is difficult to compare because emissivity values in the literature

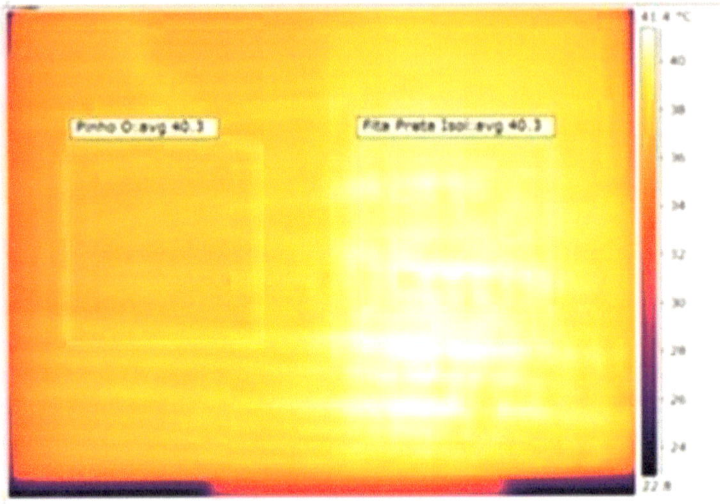

Figure 8.
Example of a thermogram obtained from one of the wood samples.

Place	Test number	Emissivity (ε)	Sample surface temperature (°C)
IPG	1	0.855	42.0
	2	0.873	43.4
	Average	**0.864**	**42.7**

Table 2.
Emissivity values—Leiria Pine sample.

Place	Test number	Emissivity (ε)	Sample surface temperature (°C)
IPG	1	0.855	47.1
	2	0.890	43.0
	Average	**0.873**	**45.3**

Table 3.
Emissivity values—Serra da Estrela Pine sample.

Samples	Number of tests	Volume (cm^3)	Dry weight (g)	Weight (g)	Water content (%)	Density
Leiria pine	2	4500	2507	2810	12.1	0.624
Serra da Estrela pine	2	4500	1869	2058	10.1	0.457

Table 4.
Samples conditions and emissivity.

refer to test conditions and experiments not fully described. The environmental characteristics to be described are ambient temperature and the relative humidity. The spectral observation window must be referred to. The conditions that refer to the sample itself, such as surface finishing, water content, density and the wood species itself, must also be mentioned.

The wood has values of emissivity, very dependent on the spectral window used.

When referring to emissivity of a natural material, such as wood, it is important to describe the set of factors conditioning the emissivity: the distance between the machine and the sample, observation angle, spectral window for observation, ambient temperature, and other ambient conditions such as relative humidity, light intensity, reflected temperature and sample surface temperature.

The difference between the sample surface temperature and the ambient temperature must be as constant as possible during the determination of the emissivity. According to the ASTM standard [19], the higher the difference the more rigorous will be the emissivity determination. The upper limit is the temperature value at which the physical properties of the object of study deteriorate.

The type of surface finishing, species identification (by scientific name), colour, sample water content and density should also be described. All these factors condition the emissivity measurement. On the other hand, only this way, by referring all the parameters, make it possible to obtain a comparative notion of the emissivity determined under different conditions. Therefore, more correctly will be the choice, for a real situation of surface temperature analysis with IR thermography technique. Even so, variations are expected in the resulting values because of both sets of environmental and sample parameters. In fact, when referring to sample parameters, there is a large variability among wood species, besides that, there is large variability along the tree itself from which the sample is taken (heartwood or sapwood). Even more, variation occurs depending upon the cutting technique (parallel or perpendicular direction to the fibre).

When reviewing the literature, it was not found systematised emissivity values under the same conditions, that is, for example, ambient temperature, spectral window and species under study. The literature on the subject is scarce and generally does not present all the relevant parameters to be taken into account for measurement.

In addition, it is not common the authors to describe the process applied to heat the surface of the sample although this is another factor that conditions the emissivity value [9]. It is relevant because not all types of heating are suitable for this purpose [18].

5. Conclusion

The aim of this research is to contribute to the obtainment of information relevant to the investigation of wooden building conservation. It was carried out under an experimental approach using pine wood of the *Pinus pinaster* species.

Emissivity values were obtained from two different samples of the same wood species. Emissivity values were obtained at an ambient temperature of 21.0°C, at a distance of 0.5 m. The sample dimensions of the observed face were 0.20 × 0.15 m to minimise errors from the surroundings. The spectral band used was the one the FLIR B20 thermographic camera provides, that is 7.5–13 μm. The samples did not have any type of finish. Lab conditions intended to reproduce the *in situ* conditions. The specimens studied, although of the same species, came from different climate zones and had different density values.

The emissivity values obtained are into the emissivity range suggested in the literature. However, it was not found literature that met the same conditions of observation regarding the species, ambient temperature, relative humidity and the spectral band used in this experimental work.

Experiments of this type are relevant since an incorrect emissivity measurement can lead to inaccurate results in the interpretation of the thermograms and hence to false conclusions.

The correct way to obtain temperature values at IR thermography systems is to establish the emissivity of the materials to be tested. Nevertheless, it is often not possible in the course of *in situ* investigations. In that case, samples of material should then be collected and tested in the laboratory, reproducing carefully the same environmental conditions as those found *in situ* to avoid distortions that may bias the results.

In the literature, we found few published works on emissivity values for wood materials. Thus, a listing of wood emissivity values at different ambient temperatures for buildings/timber structures is timely and relevant.

Acknowledgements

This research is framed in the project "TreeM – Advanced Monitoring & Maintenance of Trees" N. ° 023831, 02/SAICT/2016, co-financed by CENTRO 2020 and FCT, Portugal 2020 and structural funds UE-FEDER.

Conflict of interest

The authors confirm that there are no conflicts of interest.

Notes/Thanks/Other declarations

The authors are grateful for the kindness of Professor Luis Jorge, from the Polytechnic Institute of Castelo Branco, Portugal, for supplying the *Pinus pinaster* sample from the coastal region of Leiria, Portugal.

Author details

João Crisóstomo and Rui Pitarma*
Polytechnic Institute of Guarda, Guarda, Portugal

*Address all correspondence to: rpitarma@ipg.pt

References

[1] Wyckhuyse A, Maldaque X. Wood inspection by infrared thermography. In: Proceedings of IVth IWASPNDE; 2002. Quebec, Canada

[2] Bucur V. Techniques for high resolution imaging of wood structure: A review. Measurement Science and Technology. 2003;**14**:R91-R98. DOI: 10.1088/0957-0233/14/12/R01

[3] Bucur V. Nondestructive Characterization and Imaging of Wood. New York: Springer; 2003. pp. 75-77. DOI: 10.1007/978-3-662-08986-6

[4] Ciocia C, Marinetti S. In-situ emissivity measurement of construction materials. In: Proceedings of 11th International Conference on Quantitative Infrared Thermography; 2012. Naples, Italy

[5] Maldague X. Nondestructive Evaluation of Materials by Infrared Thermography. London: Springer-Verlag; 1993. 139 p. DOI: 10.1007/978-1-4471-1995-1

[6] Rosina E, Robison E. The application of infrared thermography on wood-framed North American historic buildings. APT Bulletin. 2002;**4**:37-44

[7] Crisóstomo J. Contribuição para a Análise de Manifestações Patológicas em Madeira na Construção com Recurso à Termografia – Determinação da Emissividade de Algumas Espécies de Madeira Empregues em Portugal [thesis]. Castelo Branco: Superior School of Technology of the Polytechnic Institute of Castelo Branco; 2014

[8] Rice R. Emittance factors for infrared thermometers used for wood products. Wood and Fiber Science. 2004;**36**:520-526

[9] Fronapfel E, Stolz B. Emissivity Measurements of Common Construction Materials [Internet]. Available from: http://www.callsbsa.com/wp-content/uploads/2011/03/Emissivity-Measurements-of-Common-Construction-Materials.pdf [Accessed: January 21, 2013]

[10] Avdelidis N, Moropoulou A. Emissivity considerations in building thermography. Energy and Buildings. 2003;**35**:663-667. DOI: 10.1016/S0378-7788(02)00210-4

[11] Marinetti S, Sesaratto P. Emissivity estimation for accurate quantitative thermography. NDT and E International. 2012;**51**:127-134

[12] Wild W. Application of infrared thermography in civil engineering. In: Proceedings of 13th Estonian Academy of Sciences Engineering; 2007

[13] Lombillo I, Agudo E, Villegas L. Evaluación no Destructiva Del Patrimonio Construido: Inspección Visual, Técnicas Acústicas, Electromagnéticas y Tomográficas [Internet]. Available from: http://grupos.unican.es/gted/archivos/otros/ponencias/aend.pdf [Accessed: September 21, 2018]

[14] Linán C, Conde M, Rubio de Hita P, Gálvez F. Inspección mediante técnicas no destructivas de un edificio histórico: Oratorio San Felipe Neri (Cadiz). Informes de la Construcción. 2011;**63**:13-22. DOI: 10.3989/ic.10.032

[15] Kandemir-Yucel A, Tavukcuoglu A, Caner-Saltik E. In situ assessment of structural timber elements of a historic building by infrared thermography and ultrasonic velocity. Infrared Physics & Technology. 2007;**49**:243-248. DOI: 10.1016/j.infrared.2006.06.012

[16] Tanaka T, Divós F. Wood Inspection by Thermography [Internet]. Available from: https://www.ndt.net/article/v06n03/tanaka/tanaka.htm [Accessed: September 21, 2018]

[17] López G, Basterra L, Acuna L, Casado M. Determination of the emissivity of wood for inspection by infrared thermography. Journal of Nondestructive Evaluation. 2013;**32**(2):172-176. DOI: 10.1007/s10921-013-0170-3

[18] FLIR Systems. FLIR ThermaCAM B20 User's Manual. FLIR; 2006. 175-176 pp.; 209-210 pp

[19] ASTM E1933-99a (Reapproved 2010), Standard Test Methods for Measuring and Compensating for Emissivity Using Infrared Imaging Radiometers, American Society for Testing and Materials; West Conshohocken, USA; 2013

[20] Catena A, Catena G. Overview of thermal imaging for tree assessment. Arboricultural Journal. 2008;**30**:259-270. DOI: 10.1080/03071375.2008.9747505

[21] Bellett-Travers M, Morris S. The relationship between surface temperature and radial wood thickness of twelve trees harvested in notttinghamshire. Arboricultural Journal. 2010;**33**(1):15-26. DOI: 10.1080/03071375.2010.9747589

[22] Crisóstomo J, Pereira C, Roque E, Jorge L, Pitarma R. Considerações na observação do estado de salubridade de Árvores através da termografia por infravermelhos. In: Proceedings of the 1st Iberic Conference on Theoretical and Experimental Mechanics and Materials/11th National Congress on Experimental Mechanics; 4-7 November 2018; Porto. pp. 745-748

[23] Crisóstomo J, Pereira C, Roque E, Jorge L, Pitarma R. Análise da salubridade de Árvores através da termografia por infravermelhos. In: Proceedings of the 1st Iberic Conference on Theoretical and Experimental Mechanics and Materials/11th National Congress on Experimental Mechanics; 4-7 November 2018; Porto. pp. 749-758

[24] Grossman J. IR Thermography as a Tool for the Pest Management Professional. Miami, Florida: Hi-Tech Inspection Services, Inc. [Internet]; 2004. Available from: https://www.irinfo.org/articleofmonth/pdf/article_8_1_2004_grossman.pdf [Accessed: November 20, 2018]

[25] Grossman J. Trestles Anyone?: A Thermographic Nightmare, Miami, Florida, Hi-Tech Inspection Services, Inc. [Internet]. 2006. Available from: https://www.irinfo.org/articleofmonth/pdf/article_6_2006_grossman.pdf [Accessed: November 20, 2018]

[26] Kang W, Lee YH, Kang CW, Chung WY, Xu HL, Matsumura J. Using the inverse method to estimate the solar absorptivity and emissivity of wood exposed to the outdoor environment. Journal of the Faculty of Agriculture, Kyushu University. 2011;**56**(1):139-148

[27] FLIR Systems. FLIR ThermaCAM QuickView 1.3 Software User's Manual, FLIR; 2006

[28] FLIR Systems. FLIR ThermaCAM Reporter 7.0 Software User's Manual, FLIR; 2006

[29] Spencer M, Penn L, Nychka J, Boyer L, Liebertz J, Everdale L. Applying Infrared Thermography for the Purpose of Identifying Concealed Wood Framing Member Type and Subsurface Anomalies with Intended Application Towards Historic Structures [Internet]. Avalaible from: http://ncptt.nps.gov/wp-content/uploads/2008-06.pdf [Accessed: September 21, 2018]

[30] IPQ, NP 614: Madeiras – Determinação do Teor em Água; 1973

[31] Holst G. Common Sense Approach to Thermal Imaging. Washington: SPIE Optical Engineering Press; 2000. Monograph Vol. PM86

Chapter 4

Applications of Infrared Thermography for Non-destructive Characterization of Concrete Structures

Ravibabu Mulaveesala, Geetika Dua and Vanita Arora

Abstract

Usage of reinforced concrete structures has very long tradition in infrastructure industry due to their low cost, high strength, robustness, sustainability along with the easy availability of raw materials. However, they also have some drawbacks such as poor tensile strength and ductility, which leads to the formation of cracks in the structures. These cracks may cause penetration of chlorides, resulting into corrosion in the reinforcement. Quality control, maintenance and planning for the restoration of these structures demands a suitable non-destructive testing and evaluation method for wide-area monitoring to detect the hidden corrosion of the rebar at an early stage. Infrared thermal wave imaging has emerged as a viable technique for non-destructive testing and evaluation of reinforced concrete structures due to its full-field, remote, fast inspection capabilities to monitor the sub-surface rebar corrosion. Among the various thermal non-destructive testing techniques the present chapter proposes a novel aperiodic thermal wave imaging technique named as Gaussian windowed frequency modulated thermal wave imaging for testing and evaluation of rebar corrosion in concrete structures.

Keywords: concrete structures, frequency modulation, image processing, matched filter, non-destructive testing, phase images, pulse compression, rebar, thermal non-destructive testing, thermal wave imaging

1. Introduction

Reinforced concrete is a durable material, capable of bearing various severe environmental conditions. It is an extensively used material for constructing bridges, buildings, flyovers and underground structures. Despite of the fact that majority of these structures have high strength, durability and show good long term performance, there are large number of failures of concrete structures as a result of premature corrosion in reinforcement. The main reasons behind the corrosion initiation of reinforcing steel are the ingress of chloride ions and carbon dioxide to the steel surface, leading to the significant loss in cross-section. During the corrosion process, iron oxides and hydroxides being the corrosion products are usually get deposited in the concrete around the steel material. Their formation within this confined space sets up stress, which crack and spall the concrete cover. This, in turn, results in progressive deterioration of the concrete. As a result, repair costs constitute a major part of spending on infrastructure. There have been a large number of investigations on the problems of deterioration of concrete and the consequent corrosion of rebar. It is essential to monitor the state of such deliberate structures right from the construction stage by carrying out periodic inspections and thus maintaining record of the results. For the estimation of corrosion rate of reinforced steel in concrete, many electrochemical and non-destructive techniques are available.

Non-destructive testing and evaluation (NDT&E) is defined as the application of an inspection method to assess the integrity of the object without impairing its future usefulness [1]. Non-destructive testing (NDT) plays a crucial role in various applications to inspect the quality and safety of the products in a reliable and cost effective manner [1–12].

Widely used NDT&E methods can be broadly classified based on their applicability as point-wise or whole-field techniques. In point-wise methods, the test specimen is scanned on each and every point in order to reveal information regarding the hidden defects, whereas, whole-field methods provide full information of the test specimen in a single run. Depending upon the way, the test is going to be implemented i.e. the specimen to be tested is in contact with the test unit or it can be tested remotely, these NDT&E techniques can be sub-classified as contact or non-contact techniques. Further, depending on the capability of the technique to be adopted for NDT&E, these techniques can be classified as surface, sub-surface or volumetric techniques [3–20]. In view of their merits, always whole-field, non-contact and volumetric NDT&E techniques are preferable over the others. Infrared thermography (IRT) is a well-established tool in the field of non-destructive testing and evaluation [13–21].

2. Infrared thermography

Infrared thermography is based on the fact that all objects above 0 K temperature, emits infrared energy as a function of their temperature. Infrared radiation corresponds to the band of the electromagnetic spectrum that lies in between 0.74 and 1000 μm wavelength region (as shown in **Figure 1**).

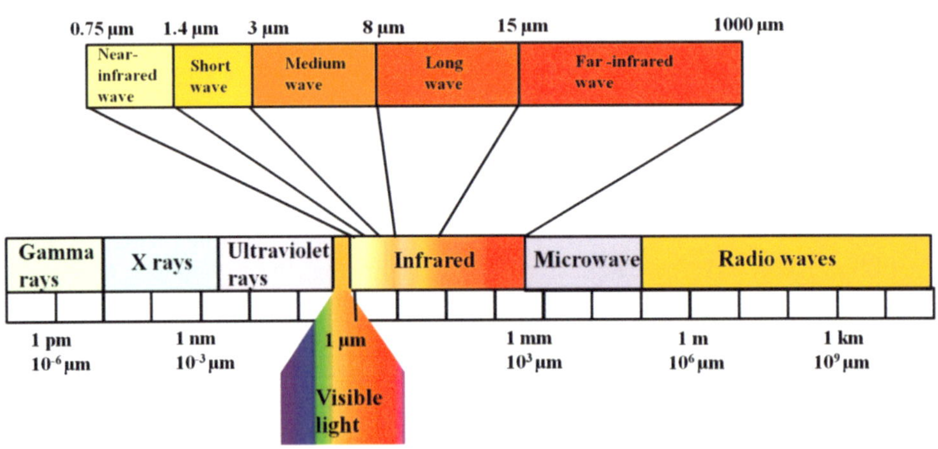

Figure 1.
Electromagnetic spectrum.

The emitted radiation at each and every wavelength and the corresponding temperature distributions for a given object is explained from the Wien's law. It states that hotter objects emit most of their radiation at shorter wavelengths whereas colder objects emit at longer wavelengths. It is as illustrated in **Figure 2**.

According to Stefan-Boltzmann law, the net radiated power per unit area is proportional to the fourth power of object's temperature as mentioned below [1]:

$$P = \varepsilon \sigma T^4 \tag{1}$$

where P is the net radiated power per unit area [W/m^2], ε is emissivity, σ is Stefan-Boltzmann constant [5.6703×10^{-8} W/m^2K^4] and T is temperature [K].

IRT can be defined as a technique which detects the infrared radiation emanating by an object, converts it into temperature and displays into an image named as thermogram. It is a fast, remote, whole-field and quantitative characterization technique which provides real-time information. It can be applied to inspect various materials and have numerous applications in various fields such as structural health monitoring (SHM), non-destructive testing, quality inspection in metal or non-metal materials such as alloys or composites, civil engineering and building sciences.

IRT for SHM/NDT&E applications involves mapping of the temperature over the structure/specimen for characterizing its surface and sub-surface structural defects. The surface of the test specimen is illuminated using heat sources and infrared camera is used to map the resultant temperature profile over the specimen. Further, the recorded temperature distribution is processed and analyzed to produce thermal images. IRT has gained wide acceptance and plays a crucial role in the field of testing and evaluation due to its merits such as non-contact, whole-field, safe and fast inspection capabilities for characterization of various materials irrespective of their electrical, optical and magnetic properties.

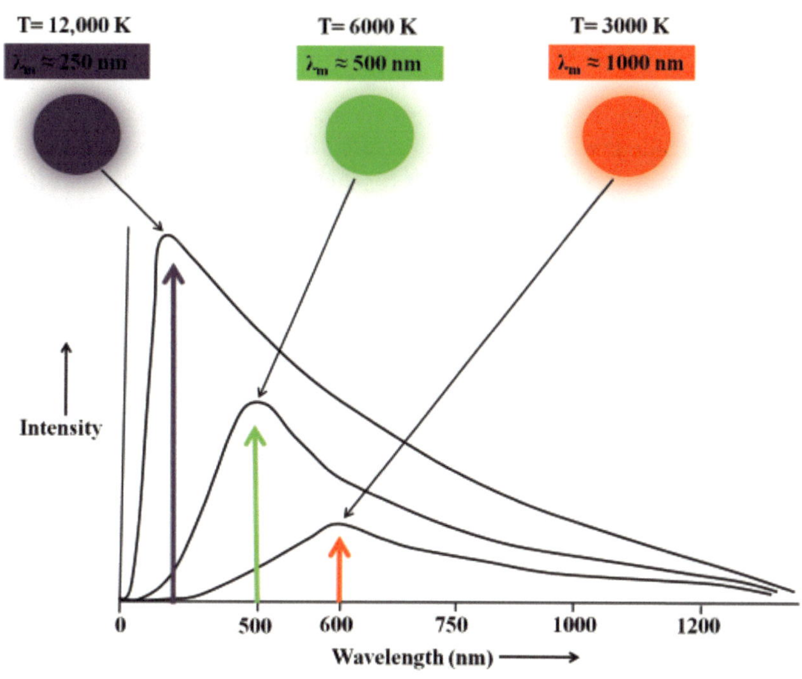

Figure 2.
Illustration of Wein's law.

IRT can be implemented in two ways: passive and active. In passive approach, the test specimen or the structure is naturally at different temperature than the ambient i.e. the passive approach is used when the defective/faulty regions in the test specimen provides enough thermal contrast with respect to the sound (non-faulty) region. Typical applications of passive thermography includes seepage/moisture assessment, insulation problems in buildings, condition monitoring, structural health monitoring, etc. However, due to the limited depth of penetration and inability in providing the quantitative assessment for the sub-surface defects deep inside the test specimen, restricts its applicability for NDT&E applications [13–21].

On contrary, to detect defects located deep inside the test specimen/stress concentration points inside the structure with enough thermal contrast, active thermography is preferred. In active approach, an external heat stimulus with a predefined amplitude, duration and bandwidth is imposed onto the test specimen. These known characteristics of the external thermal stimulus, helps in providing quantitative estimation of the sub-surface defects.

2.1 Widely used active IRT techniques for testing and evaluation of concrete structures

Active IRT (AIRT) has been developed as a non-destructive testing technique to detect surface and sub-surface defects by monitoring the emitted thermal radiations over the test object in response to a predefined applied excitation. A typical experimentation used to carry out AIRT is as shown in **Figure 3**.

AIRT is an emerging technique that has the potential to provide quantitative information about the hidden defects in a limited span of time. Based on the employed excitation scheme, AIRT can be classified as pulsed and modulated thermographic methods. Pulse thermography (PT) [3–9] and pulse phase thermography (PPT) [11, 12] are the two pulse based thermographic methods.

In PT, the test specimen is excited using a short duration high peak power pulse and the corresponding temporal temperature distribution is monitored. The main advantage of the PT approach is its fast implementation. In general, the results obtained with this technique are based on the raw temperature profile obtained over the test specimen. The obtained thermal data is not only influenced by the sub-surface defects inside the test structure but also affected by the emissivity as well as non-uniform illumination variations over it. These artifacts may lead in bringing difficulties in interpretation of test results. In addition to these, the requirement of

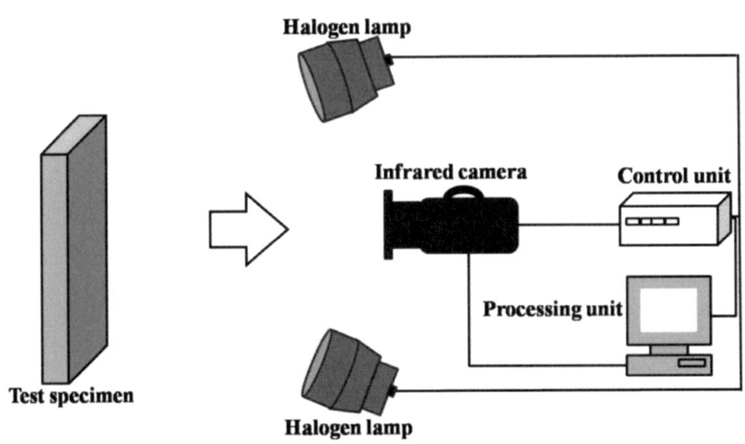

Figure 3.
Experimental set up.

high peak power heat sources to probe high frequency thermal waves into the test specimen (to resolve defects located inside the test specimen at various depths of different lateral dimensions) limits its applicability as far as the experimental setup cost is concerned.

Experimentation involved in pulse phase thermography is similar to PT (i.e. the test specimen is excited with a short duration high peak power pulse and the temperature variation over the surface of the structure is captured using an infrared camera) but it differs from the involved post-processing on the captured temporal temperature distribution. In this, the captured thermal distribution is analyzed in frequency domain using discrete Fourier transform (DFT) to extract the phase and magnitude information from its transformed temperature distribution. Phasegrams are of particular interest in non-destructive evaluation, as phase is less affected by environmental reflections, emissivity variations and non-uniform heating than that of the raw thermal data. This phase information plays a crucial role not only for qualitative representation in identifying the sub-surface defects but also helps in providing quantitative details. In addition to its fast inspection capabilities, results obtained with this technique are less sensitive to surface features of the object such as emissivity variations and non-uniform illuminations [11]. Even though, it has above mentioned merits still the requirement of high peak power heat sources limits its applicability [14].

In order to overcome the requirement of high peak power heat sources, modulated thermographic techniques are introduced by driving the heat sources at a predefined frequency, decided by the sample thickness and its thermal properties. In general, this modulation is carried out by a single frequency to provide thermal stimulus onto the test specimen and is named as lock-in thermography (LT) [10]. In LT, a sinusoidal modulated heat flux is imposed onto the test specimen and the resultant temporal temperature distribution is captured during the active heating. Further, from the extracted temporal temperature response, the phase-grams are reconstructed either by implementing quadrature phase shift or by Fourier Transform approach. These reconstructed phase images not only have the advantages such as less sensitive to emissivity variations and non-uniform heating over the test specimen but also helps in deeper depth of probing inside the test specimen.

Even though, the technique allows detection of defects located deep inside the test specimen with moderate peak power heat sources, but the mono-frequency excitation limits its applicability in identifying defects located at various depths inside the test specimen with enough resolution. In order to detect defects located at different depths of various lateral dimensions with enough resolution, LT has to be carried out with different excitation frequencies. This makes LT a time-consuming testing technique. These limitations can be overcome by probing thermal waves with suitable band of frequencies having significant magnitude into the test sample in a single experimentation cycle with relatively low peak power sources. To achieve this, the present chapter focuses on one of the available aperiodic thermal wave imaging techniques i.e. linear frequency modulated thermal wave imaging. This technique is introduced to overcome the limitations (resolution, peak power, limited depth of penetration) of the conventional (PT, PPT, LT) thermographic techniques.

2.2 Gaussian windowed frequency modulated thermal wave imaging

This present chapter highlights the capabilities of aperiodic linear frequency modulated thermal wave imaging (LFMTWI) and the associated post-processing techniques for detection of corrosion in rebar [18–21]. In LFMTWI, the surface of the test specimen is heated by driving the heat sources with a linear frequency modulated heat flux, which results into a similar frequency modulated temperature distribution over it, with a mean rise in temperature depending on the total duration of the excitation. This helps in probing the desired band of thermal waves (decided by thermal properties of the test specimen, location and spatial dimensions of the defects) with significant magnitude in a limited time span into the test specimen which improves defect detection resolution.

However, smaller ratio of the concentration of the supplied energy in the main lobe to that of side lobe levels of the compressed pulse obtained from the matched filtering processing limits its defect detection resolution and sensitivity. The present chapter highlights the concept of spectral reshaping considered for proposed LFMTWI, in order to improve the pulse compression properties which lead to enhance the defect detection resolution and sensitivity. It is incorporated by reshaping the captured temporal thermal distribution obtained from LFMTWI experimentation using a Gaussian window and is named as Gaussian windowed frequency modulated thermal wave imaging (GWFMTWI). Further, the capabilities of GWFMTWI for detection of corrosion in reinforced rebar are compared with LFMTWI considering signal to noise ratio (SNR) as a figure of merit.

3. Theory

In LFMTWI, linear frequency modulated (LFM) heat flux is imposed onto the test sample to probe thermal waves into it. These thermal waves diffuse into the test sample by producing similar time varying temperature distributions over the surface except a mean rise during the dynamic heating. The theoretical model represented to study this thermal response can be derived from the one-dimensional heat diffusion equation given as [14]:

$$\frac{\partial^2 T(x',t')}{\partial^2 x'} - \frac{1}{\alpha}\frac{\partial T(x',t')}{\partial t'} = 0 \tag{2}$$

where α is the thermal diffusivity ($\alpha = k'/\rho.c$); ρ, c, k' are the density, specific heat and thermal conductivity of the medium respectively. $T(x',t')$ is the temperature response over the surface and x' is the spatial dimension in the direction of heat flow.

The proposed (frequency modulated) heat flux$Q(x',t')$ of duration τ [s], with a bandwidth B [Hz], is obtained by means of the LFM excitation signal used to drive the heat sources [14],

$$Q(x'=0,t') = Q_0 e^{2\pi j\left(f_0 t' + \frac{Bt'^2}{2\tau}\right)} \tag{3}$$

where Q_0 is the envelope of the frequency modulated incident heat flux (chirp signal), which is zero outside the time interval τ [s], f_0 is the initial frequency [Hz] and $2\pi j\left(f_0 t' + \frac{Bt'^2}{2\tau}\right)$ is the phase of the LFM (chirp) incident heat flux.

The solution to heat equation (Eq. (2)) for a LFM heat flux (Eq. (3)) over the test specimen by considering it as a semi-infinite solid with specified boundary ($x' = 0$, a similar kind of temperature response is expected and $x' \to \infty$, ambient temperature) and initial conditions ($T(x', t' = 0) = 0$)) is obtained as [14]:

$$T(x',t') = T_0 e^{2\pi j\left(f_0 t' + \frac{Bt'^2}{2\tau}\right)} e^{-x'\sqrt{\frac{\pi}{\alpha}\left(f_0 + \frac{Bt'}{\tau}\right)}} e^{-jx'\sqrt{\frac{\pi}{\alpha}\left(f_0 + \frac{Bt'}{\tau}\right)}} - \frac{2T_0}{\sqrt{\pi}} e^{2\pi j\left(f_0 t' + \frac{Bt'^2}{2\tau}\right)} \int_0^{x'/2\sqrt{\alpha t'}} e^{\frac{-\pi j x'^2}{2\alpha \mu^2}\left(f_0 + \frac{Bt'}{\tau}\right)} e^{-\mu^2} d\mu \tag{4}$$

The second term in Eq. (4) is a transient disturbance caused by starting oscillations of surface temperature at initial time. It dies away as t' increases, leaving the first term which is a steady-state solution i.e.

$$T(x',t') = T_0 e^{2\pi j\left(f_0 t' + \frac{Bt'^2}{2\tau}\right)} e^{-x'\sqrt{\frac{\pi}{\alpha}\left(f_0 + \frac{Bt'}{\tau}\right)}} e^{-jx'\sqrt{\frac{\pi}{\alpha}\left(f_0 + \frac{Bt'}{\tau}\right)}} \tag{5}$$

where B/τ is the modulation factor and τ is duration of excitation. The thermal diffusion length μ [m] from the above solution (Eq. (5)) can be derived as [14]:

$$\mu = \sqrt{\frac{\alpha}{\pi\left(f_0 + \frac{Bt'}{\tau}\right)}} \tag{6}$$

Thus, it can be justified from Eq. (6) that LFMTWI ensures depth scanning of the sample with time in a single experimentation cycle.

In GWFMTWI technique, the envelope of the obtained solution for LFMTWI (Eq. (5)) is reshaped by using Gaussian window function given as in Eq. (7):

$$g(t') = e^{-\frac{(t'-\mu')^2}{2(\sigma')^2}} \tag{7}$$

where μ' and σ' are the mean and standard deviation respectively.

To facilitate the spectral reshaping on LFMTWI, the Fourier transform (FT) of $g(t')$,$(G(f))$ is multiplied with FT of the captured temporal temperature distribution at a given location ($T(x_i',f)$) over the test specimen and then converting it back to time domain using inverse Fourier transform (IFT). The obtained Gaussian windowed chirp (GWC) response (TGWC) can be represented as in Eq. (8):

$$T_{GWC/Gaussian} = \text{IFT}(G(f).T(x_i',f)) \tag{8}$$

4. Results and discussion

To validate the proposed approach, a concrete sample of 6.7 cm thickness, containing 4 cm thickness mild steel rebar of length 13.24 cm is considered. The rebar is placed in sample by wounding a cotton cloth to fill up the groves in order to avoid incursion of concrete into the grooved area. A simulated corrosion has been introduced to the rebar by introducing four grove cuts of different widths (a, b, c and d) with a material loss of 5 mm from the surface of bar. The front view, side view and its experimental orientation is as shown in **Figure 4**. Further, experiments have been performed with LFMTWI technique on this sample. The experimental setup is shown in **Figure 5**.

A linear frequency modulated (LFM) signal with a frequency variation from 0.01 to 0.1 Hz at a sampling rate of 25 Hz of 100 s duration, is generated to drive the heat sources (two halogen lamps of each 1000 W/m^2) by the power control unit. During the active heating over the test structure for a frequency modulated incident heat flux, temporal temperature distribution over the test specimen is captured by an infrared camera (FLIR 5500) and images (thermograms) are captured at every 0.04 s time intervals for duration of 100 s. Further, temporal temperature distribution for each pixel is obtained from the thermal sequence and fitted with a linear fit. Post processing of the obtained data has been carried out after removing the mean rise of the thermal response for a given frequency modulated excitation for obtaining the dynamic variation in the temperature data. Next, Gaussian windowed (with mean 50 and standard deviation of 28) spectral reshaping have been

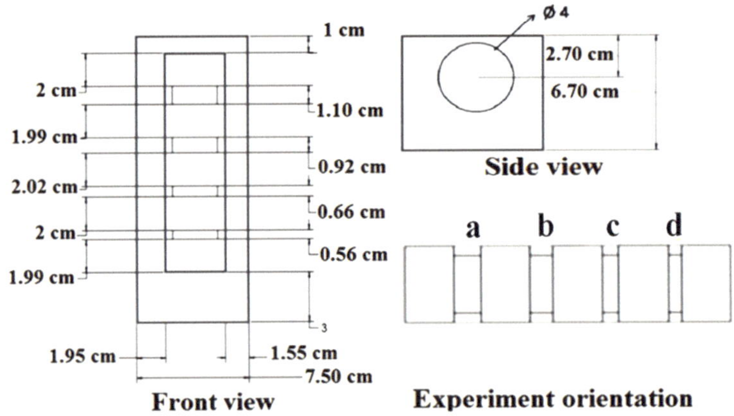

Figure 4.
Schematic of the experimental concrete sample.

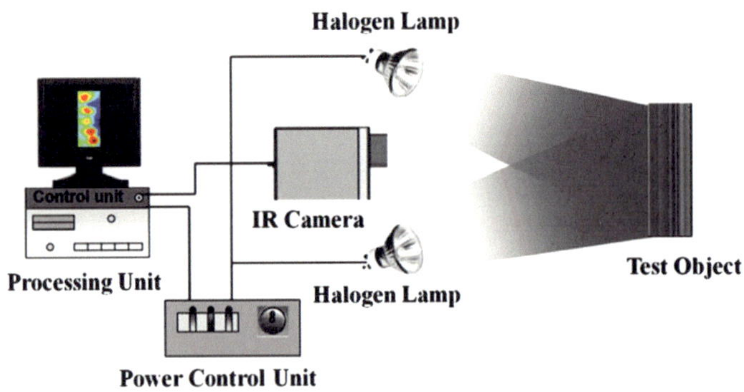

Figure 5.
Schematic of infrared imaging system used for experimentation.

performed onto the zero mean data to improve the pulse compression properties. Multi-transform techniques have been implemented both in time and frequency domain in order to compare sub-surface defect detection capabilities of these LFMTWI and GWFMTWI techniques. The conventional frequency domain amplitude (magnitude) and phase images are obtained by taking the FT of the Gaussian windowed temporal temperature data ($T_{Gaussian}(x, y, t)$), to the pixels in the field of view as below [2, 11, 19]:

$$T(x, y, k) = \sum_{n=0}^{N-1} T_{Gauusian}(x, y, t) e^{-\frac{j2\pi kn}{N}} = \mathrm{Re}(T(x, y, k)) + jImg(T(x, y, k)) \tag{9}$$

where k is the bin number, N is total number of frames, $Re(T(x,y,k))$and $Img(T(x,y,k))$ are the real and imaginary parts of $T(x,y,k)$ respectively. Further, the magnitude images are reconstructed as [2, 11, 19]:

$$|T(x, y, k)| = \sqrt{(\mathrm{Re}(T(x, y, k)))^2 + (Img(T(x, y, k)))^2} \tag{10}$$

and the phase images are reconstructed by using the equation as [2, 11, 19]:

$$\angle T(x, y, k\ ) = \tan^{-1}\left(\frac{Img(T(x, y, k))}{\mathrm{Re}(T(x, y, k))}\right) \tag{11}$$

The time domain matched filtering has been carried out in two different approaches, in the first a linear correlation approach has been implemented on the zero mean Gaussian temporal temperature distribution whereas in the later, a circular correlation approach has been adopted. The flowchart representing the pulse compression approach using linear correlation approach is as shown in **Figure 6**, whereas **Figure 7** represents the process for obtaining the pulse compression (PC) using circular correlation. Mathematically, pulse compression using linear correlation can be obtained as [13–19]:

$$T_{PC}(x,y,\tau') = \int_{-\infty}^{+\infty} T_{Gauusian}(x,y,t').T_{Reference}(x,y,(t'-\tau'))dt' \quad (12)$$

whereas, the pulse compression using circular correlation is obtained as:

$$T_{CCC}(x,y,t') = IFT\{FT(T_{Reference}(x,y,t'))^{*}.FT(T_{Gaussian}(x,y,t'))\} \quad (13)$$

The resultant images obtained after post processing are shown in **Figure 8**. The magnitude images obtained at a frequency of 0.02 Hz as a result of frequency domain analysis are as shown in **Figure 8(a)** using LFMTWI and (b) obtained for GWFMTWI. It's clear from the computed SNRs (as in **Table 1**) that the obtained frequency domain magnitude image from GWFMTWI technique is having better detection capabilities than that of the obtained frequency domain magnitude image for the FMTWI technique. Further, the reconstructed phase images at a frequency of 0.06 Hz and 6.84 Hz using LFMTWI and GWFMTWI techniques are as shown in **Figure 8(c)** and **(d)**, respectively. Unlike the facts mentioned in literature regarding the depth resolvability of the frequency domain phase images and their advantages and their merits over the magnitude images, the present studies over the concrete structures showed the magnitude images have better resolvability than that of the obtained phase grams to detect the corrosion of the rebar inside concrete structures. This is evident from the computed SNRs tabulated in **Table 1**.

The obtained pulse compressed correlation coefficient images reconstructed by the correlation approach for LFMTWI (at a time instant of 62.92 s) and GWFMTWI (at a time instant of 46.32 s) techniques are as shown in **Figure 8(e)** and **(f)** respectively. Due to the efficient pulse compression capabilities of the GWFMTWI technique, obtained linear correlation coefficient results shows the better SNR in comparison to that of LFMTWI technique. However, the time domain amplitude (correlation coefficient) images reconstructed from circular correlation approach (CCC) at a time instant of 56 s are as shown in **Figure 8(g)** and **(h)** obtained for LFMTWI and GWFMTWI techniques respectively. It is clear that the obtained results from the proposed CCC approach, proposed pulse compression approaches are far superior than that of the conventional frequency domain phase approach in order to detect the corrosion in rebar hidden inside the concrete structures.

It is observed from the results that the magnitude/correlation coefficient images shows their merits in comparison to that of the conventional frequency domain phase images in order to resolve the hidden corrosion of the rebar inside the concrete structures.

For quantitative comparison among the adopted multi-transform schemes SNR is taken as a figure of merit. The values of the SNRs are computed for the grove cuts (a, b, c and d) and the comparison has been made for various time and frequency domain multi-transform techniques with (LFM) and with spectral reshaping (GWFM) are tabulated in **Table 1** and also presented in the bar graph as shown in **Figure 9**.

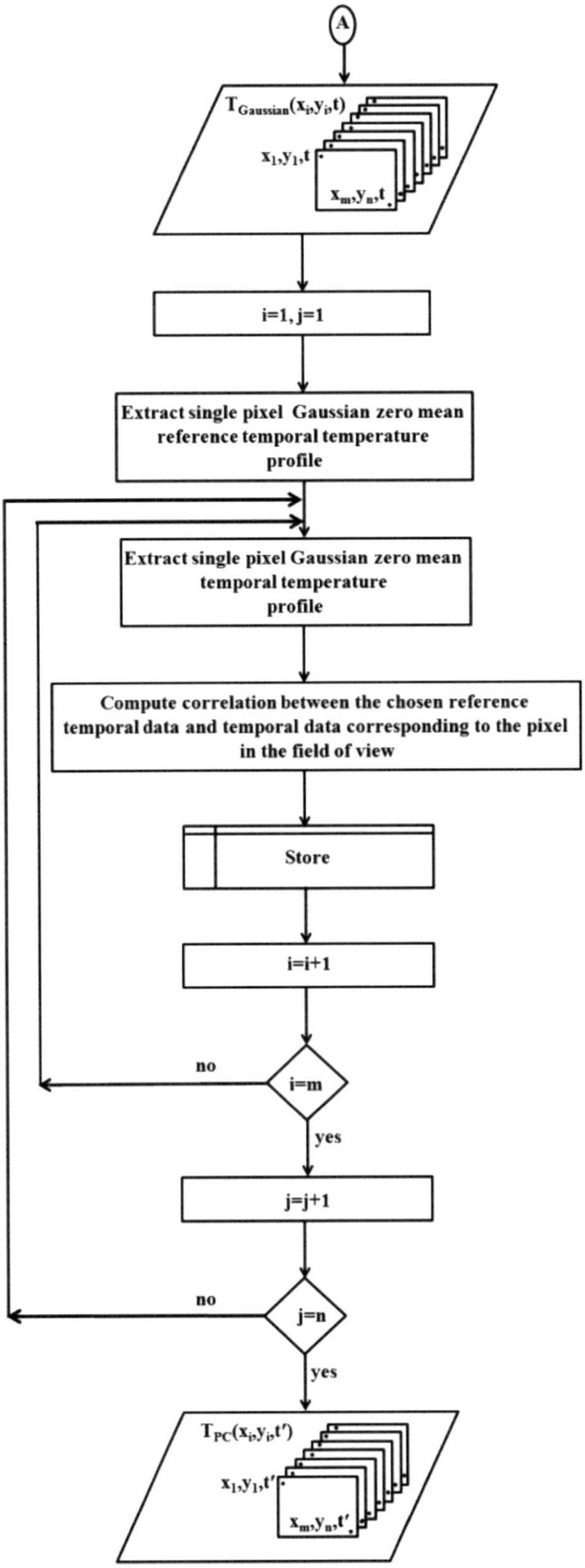

Figure 6.
Flow chart for obtaining pulse compression using linear correlation approach [19].

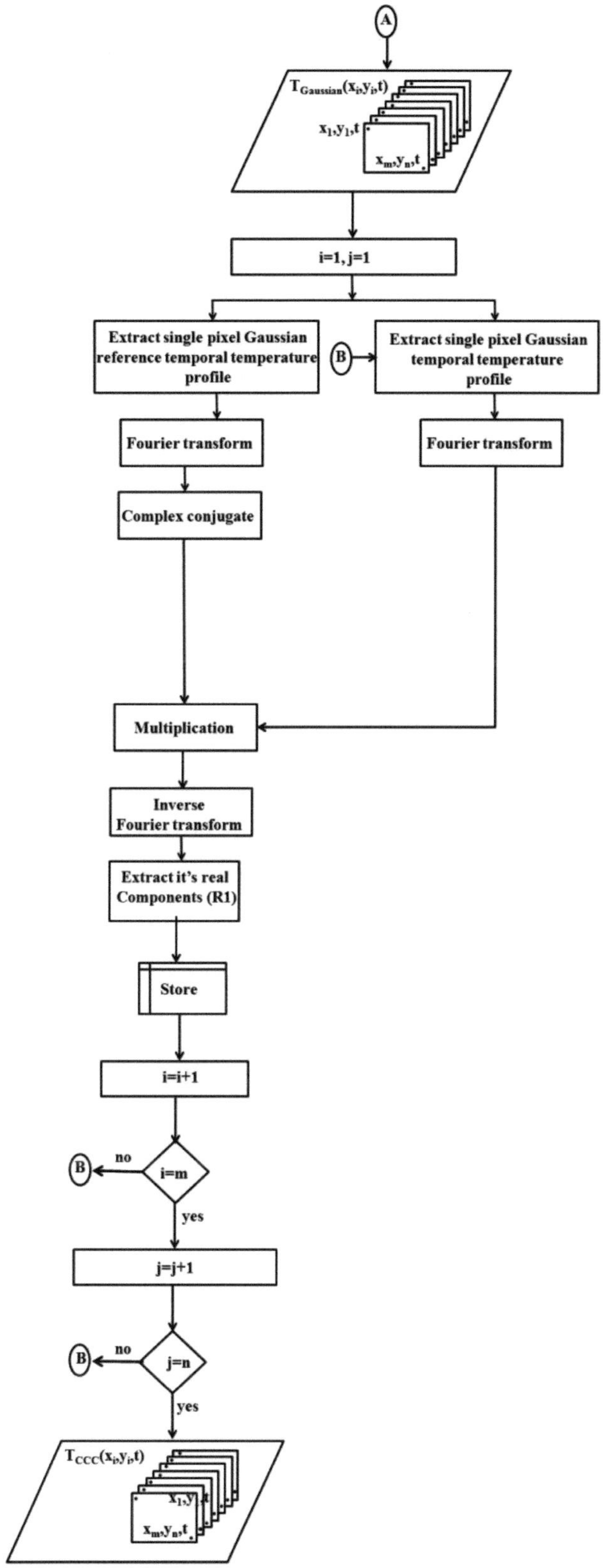

Figure 7.
Flow chart for obtaining pulse compression using circular correlation approach [19].

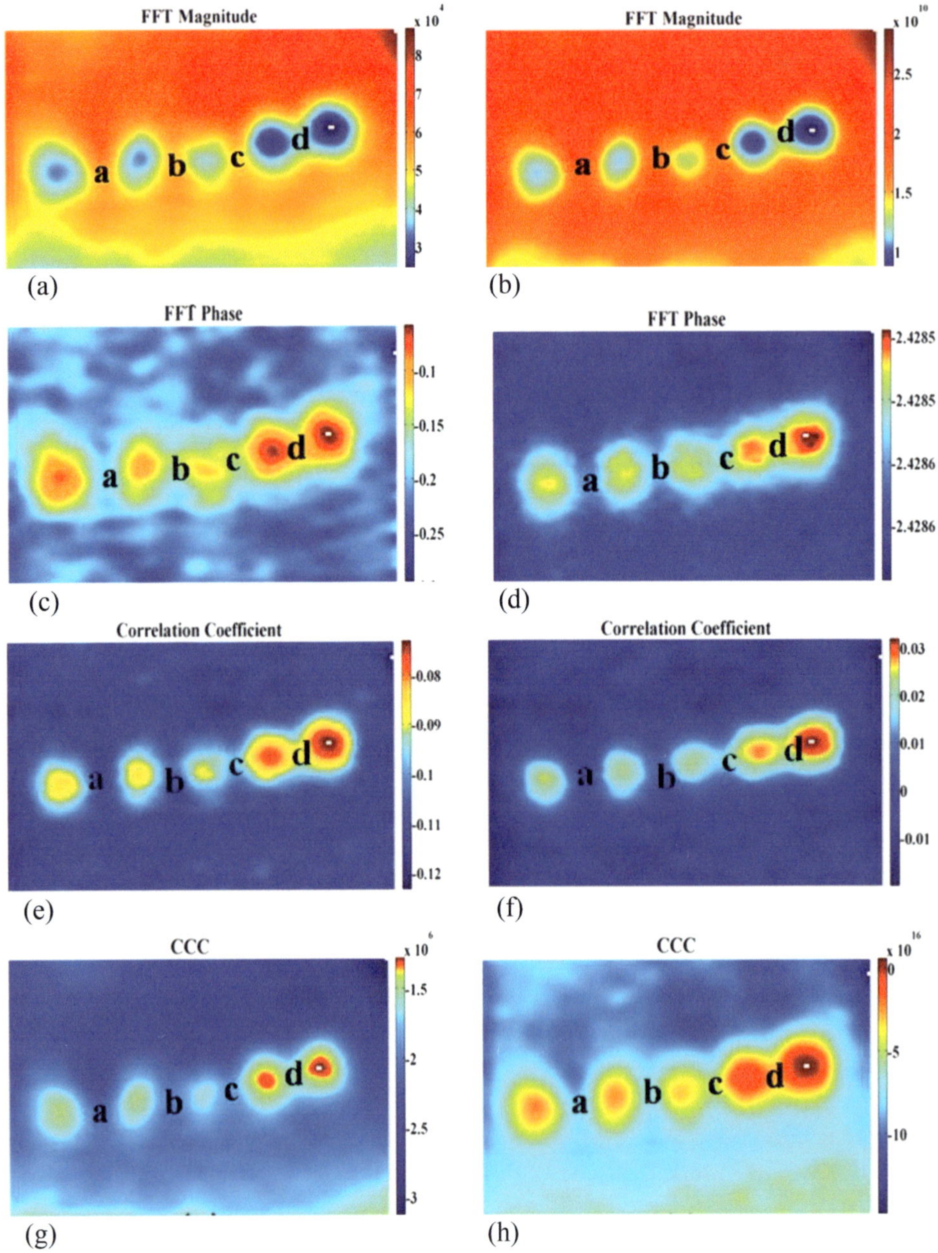

Figure 8.
Results obtained for frequency and time domain processing techniques. Frequency domain magnitude images obtained at a frequency of 0.02 Hz (a) using LFMTWI and (b) using GWFMTWI. Frequency domain phase image obtained (c) at a frequency of 0.06 Hz using LFMTWI and (d) at a frequency of 6.84 Hz using GWFMTWI. Correlation image obtained by cross correlation approach (e) at 62.92 s using LFMTWI and (f) at 46.32 s using GWFMTWI. Correlation images retrieved by circular correlation process at 56 s (g) using LFMTWI and (h) using GWFMTWI [19].

SNR is computed using formula as defined in (Eq. (14)) [13–19].

$$\mathbf{\mathit{SNR}} = 20Log\left(\frac{mean\ of\ defective\ area - mean\ of\ non\ defective\ area}{standarddeviation\ of\ non\ defective\ area}\right) \tag{14}$$

Defect	FT Amp		FT phase		Linear correlation		CCC	
	LFM	GWFM	LFM	GWFM	LFM	GWFM	LFM	GWFM
a	29.84	34.83	19.93	19.69	35.10	40.67	31.57	34.55
b	34.11	37.65	34.83	21.44	40	43.51	34.22	38.03
c	34.85	39.30	36.41	24.28	44.85	46.24	35.16	39.86
d	40.56	43.98	42.38	26.66	49.72	49.45	39.11	43.51

Table 1.
Signal to noise ratio (in dB) for different grove cuts [19].

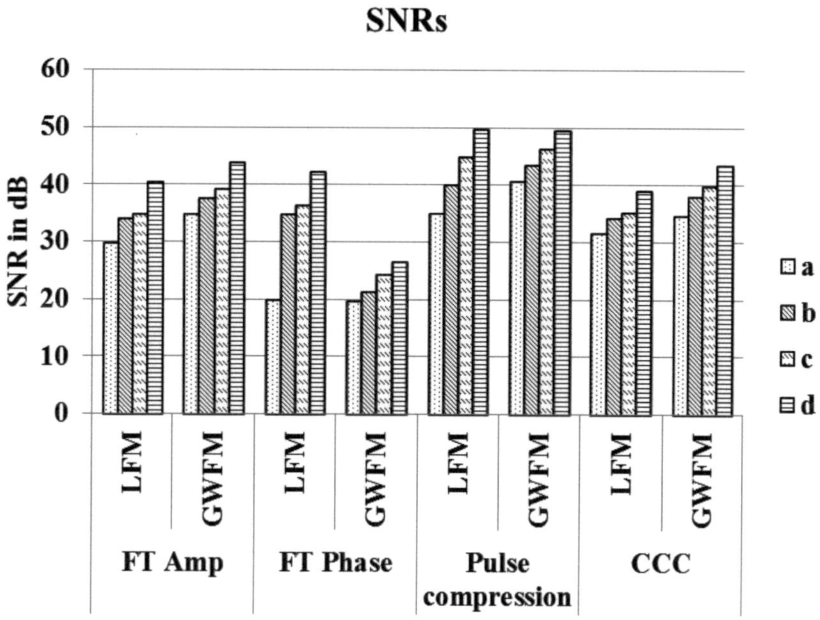

Figure 9.
Comparison of computed SNRs using different post-processing techniques for different grove cuts [19].

It is clear from **Table 1** and **Figure 9**, the results obtained using pulse compression approach shows better detection capabilities with GWFMTWI technique due to its inherent energy localization in time than that of the LFMTWI technique.

5. Conclusions

This present chapter highlights the capability of GWFMTWI technique to improve the corrosion detection capabilities of LFMTWI. Further results obtained from various multi-transform techniques in time and frequency domains are compared by considering the signal to noise ratio as a figure of merit. It clearly indicates that the GWFMTWI provides better detection performance in almost all the magnitude/correlation coefficient related data processing schemes over the conventional phase based data processing approaches. This chapter further highlights the importance of magnitude based processing approaches than that of widely used conventional frequency domain phase based approach for monitoring rebar corrosion in concrete structures.

Acknowledgements

This work was supported by the Science & Engineering Research Board (SERB), Department of Science and Technology, Govt. of India under Grant [number SB/S3/EECE/089/2014 dated 02-06-2014], and also partially supported by Aeronautics Research and Development Board (AR&DB), Govt. of India, under Grant no. DRDO/08/2031732/M/I dated 29.05.2014.

Conflict of interest

Authors have no conflict of interest.

Author details

Ravibabu Mulaveesala*, Geetika Dua and Vanita Arora
Department of Electrical Engineering, Indian Institute of Technology Ropar, Rupnagar, Punjab, India

*Address all correspondence to: ravibabucareiitd@yahoo.co.in

References

[1] Hellier C. Handbook of Nondestructive Evaluation. 2nd ed. USA: McGraw-Hill Professional Publishing; 2001

[2] Maldague X. Theory and Practice of Infrared Thermography for Nondestructive Testing. New York: Wiley; 2001

[3] Almond DP, Patel P. Photothermal Science and Techniques. Chapman and Hall Publication; 1996

[4] Mandelis A. Diffusion-Wave Fields: Mathematical Methods and Green Functions. New York: Springer; 2001

[5] Ringermacher HI, Howard DR, Filkins RJ. Flash-quenching for high resolution thermal depth imaging. Review of Progress in Quantitative Nondestructive Evaluation. 2004;**23**: 477-481

[6] Tam AC, Sullivan B. Remote sensing applications of pulsed photothermal radiometry. Applied Physics Letters. 2002;**43**:333-335

[7] Sakagami T, Kubo S. Applications of pulse heating thermography and lock-in thermography to quantitative non-destructive evaluations. Infrared Physics and Technology. 2002;**43**: 211-218

[8] Shepard SM. Introduction to active thermography for non-destructive evaluation. Anti-Corrosion Methods and Materials. 1997;**44**:236-239

[9] Almond DP, Lau SK. Defect sizing by transient thermography I: An analytical treatment. Journal of Physics D: Applied Physics. 1994;**27**:1063-1069

[10] Busse G, Wu D, Karpen W. Thermal wave imaging with phase sensitive modulated thermography. Journal of Applied Physics. 1992;**71**:3962-3965

[11] Maldague X, Galmiche F, Ziadi A. Advances in pulsed phase thermography. Infrared Physics and Technology. 2002;**43**:175-181

[12] Pickering S, Almond D. Matched excitation energy comparison of the pulse and lock-in thermography NDE techniques. NDT and E International. 2008;**41**:501-509

[13] Ghali VS, Mulaveesala R, Takei M. Frequency-modulated thermal wave imaging for non-destructive testing of carbon fiber-reinforced plastic materials. Measurement Science and Technology. 2011;**22**. art. no. 104018

[14] Mulaveesala R, Tuli S. Theory of frequency modulated thermal wave imaging for non-destructive sub-surface defect detection. Applied Physics Letters. 2006;**89**. art. no. 191913

[15] Ghali VS, Jonnalagadda N, Mulaveesala R. Three-dimensional pulse compression for infrared nondestructive testing. IEEE Sensors Journal. 2009;**9**: 832-833

[16] Tabatabaei N, Mandelis A. Thermal-wave radar: A novel subsurface imaging modality with extended depth-resolution dynamic range. Review of Scientific Instruments. 2009;**80**. art. no. 034902

[17] Ghali VS, Mulaveesala R. Comparative data processing approaches for thermal wave imaging techniques for non-destructive testing. Sensing and Imaging. 2011;**12**:15-33

[18] Mulaveesala R, Panda SSB, Mude RN, Amarnath M. Non-destructive evaluation of concrete structures by non-stationary thermalwave imaging.

Progress in Electromagnetics Research Letters. 2012;**32**:39-48

[19] Dua G, Mulaveesala R. Thermal wave imaging for non-destructive testing and evaluation of reinforced concrete structures. Insight: Non-Destructive Testing & Condition Monitoring. 2018;**60**(5):252-256

[20] Mulaveesala R, Siddiqui J, Arora V, Dua G, Subbarao GV, Muniyappa A. Testing and evaluation of concrete structures by thermal wave imaging. Proceedings of SPIE - The International Society for Optical Engineering. 2015; **9485**. art. no. 94850G

[21] Mulaveesala R, Panda SSB, Mude RN, Amarnath M. Nondestructive evaluation of concrete structures by nonstationary thermal wave imaging. Proceedings of SPIE - The International Society for Optical Engineering. 2012; **8354**. art. no. 83540B

Chapter 5

Monitoring of Critical Metallic Assets in Oil and Gas Industry Using Ultrasonic Guided Waves

Anurag Dhutti, Shehan Lowe and Tat-Hean Gan

Abstract

This chapter presents advancements in structural health monitoring (SHM) using ultrasonic guided waves (UGW) technology for metallic structures to support their integrity and maintenance management. The focus is on pipelines and storage tanks, which are critical assets in the Oil and Gas industry, whose operational conditions can greatly accelerate damage mechanisms. Conventional routine inspections are both costly and time consuming and affect the plant reliability and availability. These operational and economic disadvantages have led to development of SHM systems which can be permanently installed on these critical structures to provide information about developing damage and optimise maintenance planning and ensure structural integrity. These technology advancements enable inspection without interruption to operations, and generate diagnosis and prognosis data for condition-based maintenance, hence increasing safety and operational efficiency. The fundamentals, architecture and development of such SHM systems for pipes and above ground storage tanks are described here.

Keywords: ultrasonic guided waves, structural health monitoring, permanently installed, monitoring data analysis, defect detection, optimised maintenance planning

1. Introduction

Petroleum oil refining is an essential industry and an important element of the economic infrastructure. Refineries are large compared to other industrial plants because their production and storage capacities are designed to assure volume profitability. The industry deals with considerable amounts of flammable and toxic substances and is thus inherently hazardous. If loss of containment is not prevented or controlled, it can have serious economic and environmental consequences. The reduction of accidents is driving the development of better control technologies and risk management strategies. Corrosion remains one of the challenges which is further elevated because of ageing infrastructure and variation in concentration of crude oil.

According to a report from eMARS (Major Accident Reporting System) [1], corrosion of equipment is an important source of accidents in refineries, being responsible for one in five major refinery accidents occurring in the EU since 1984. The magnitude of a refinery unit and the complexity of the processes are great and a wide variety of equipment such as trays, drums and towers are subject to corrosion problems. The pipeline infrastructure and storage tanks are particularly vulnerable and have high risk profiles due to the volumes they may contain. The same report analysed 99 corrosion failures, 71% of them originated in pipe works and 15% of them occurred in storage tanks.

Pipelines serve as basic components of refinery infrastructure as well as the chief transmission line between refineries and remote sites delivering the products to distribution points and customers. They are generally constructed from a variant of carbon steel and so are naturally susceptible to corrosion. The intense temperatures and temperature fluctuations, and presence of corrosive agents can accelerate the corrosion process. Corrosion can cause oil leaks which may lead to explosion with severe consequences. One example is an underground oil pipeline operated by Sinopec, China's largest oil refiner [2], which exploded following an oil leak due to corrosion. The blast killed 44 people and injured 136, and led to disruption in electricity and water supply and evacuation of around 18,000 people.

Failure of storage tanks is not as prevalent as pipe work failures but due to the hazardous substances stored, they are well represented in major accidents in the process industries. Storage tanks are extensively used in refineries to store fossil fuel, acids, solvents, benzene, sour water, asphalt and related products (heated storage). Both types of storage tanks are vulnerable to corrosion. Crude oil storage tanks suffer more aggressive corrosion compared to other refinery equipment due to the oil sulphur content. Another study on storage tank accidents [3] showed that 74% of accidents involving them occurred in Petro-chemical refineries with 85% of the accidents leading to fire and explosions. One such incident happened at a fuel storage facility in Brazil in 2015 [4] which took more than 4 days to bring under control with 110 firefighters, road blockages and the shut-down of ports (**Figure 1**).

Over the years, numerous non-destructive testing (NDT) techniques have been used to inspect the condition of pipelines and storage tanks, e.g. penetrant testing, magnetic particle testing, radiography, eddy current, thermography, acoustic emission and conventional ultrasonic testing [5]. Many of these techniques only offer localised inspection. Pipe inspection using these techniques requires removal of insulation to access pipe surfaces and may even require erection of scaffolding for difficult-to-access locations. For storage tanks, exterior corrosion, whether general or localised at crevices, is easy to detect using the aforementioned inspection methods. But for inspection of internal tank floors from exposure to corrosive agents in the product, requires the tank to be emptied and cleaned to gain access. These operations are both time-consuming and expensive and cannot be used in-service.

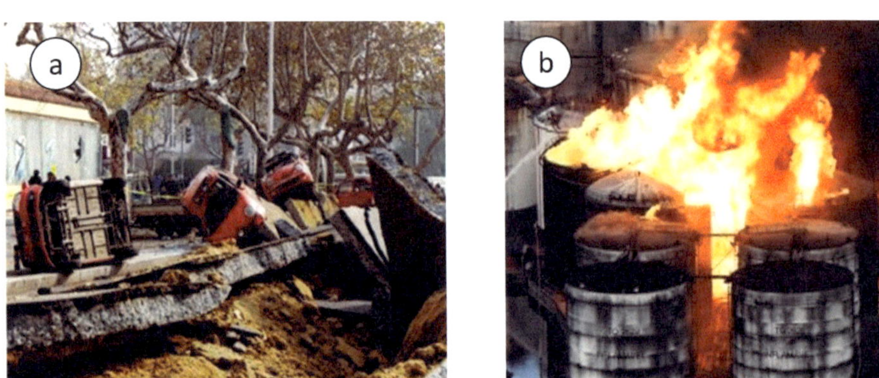

Figure 1.
The damage from (a) oil leakage of a corroded buried pipeline in China [2] and (b) tank at a fuel storage facility in Brazil [4], which led to explosions with severe consequences and put human in danger.

Less than rigorous inspection is considered a major cause of corrosion failure [1]. For this reason, there has been increased emphasis on the development of damage prognosis systems that inform the operator of a structure's health and of any developing damage. This will enable accurate estimation of the remaining useful life of the structures and can transform maintenance procedures from schedule-driven to condition-based implementation. These systems will significantly decrease the time these structures are offline, hence cutting life-cycle costs and labour requirements. Structural Health Monitoring (SHM) serves an essential part of any damage prognosis system. It monitors the structures whilst they are in-service and provides information about any detected damage.

The integration of Guided Wave Testing (GWT) technology into SHM is growing rapidly as it offers a remote solution with the ability to screen large structures. This chapter will detail the advances in SHM technologies using GWT for the two most critical metallic components in the Oil & Gas industry: pipelines and storage tanks. A brief description of GWT and the underlying physics of Ultrasonic Guided Waves (UGW) for tubular and plate like structures is provided. Its application to SHM of pipelines and storage tanks is described and the state-of-the-art in the enabling technologies including transducers and their coupling (transducer system) and data processing is presented. The design, operation and performance of SHM devices for pipelines and storage tanks are presented, and their current limitations are highlighted to direct future research and development activities.

2. Background of guided wave technology

Much research has been conducted on the use of UGWs to inspect elongated engineering structures, i.e. pipes, plates, rails and cables, because of their inherent long range propagation [6]. Commercial GWT systems have evolved vastly over the past two decades to fulfil many industrial inspection requirements. For pipes, initial realisation of UGW propagation in cylindrical structures by Gazis et al. [7], Zemanek [8] and Silk and Bainton [9], led to initial development of a GWT system [10–12] for pipes which were commercialised [13, 14] and rapidly adopted by the Oil and Gas industry. Worlton [15] and Viktorov [16] originally explored the potential of UGW for NDT of plate-like structures. Based on this, Mažeika et al. [17] studied the potential for GWT of tank floors.

2.1 Ultrasonic guided waves

Rayleigh waves [16] are surface waves that exist in half-space, a surface backed by a semi-infinite volume. These waves have an elliptical vibration with the major axis of vibration perpendicular to the direction of propagation. They can penetrate to a depth of 1.5λ below the surface. In contrast, Lamb waves fill the entire volume of the plate provided its thickness is less than 2λ. These waves were first analysed on plates by Horace Lamb [18] and can be considered as Rayleigh waves bounded by two parallel surfaces. In plates, there are three fundamental wave modes in the operating frequency range for GWT: namely, the fundamental Symmetric Lamb mode, S0, the Asymmetric Lamb mode, A0, and the Shear Horizontal (SH) mode, SH0, as illustrated in **Figure 2**.

Just like plates, hollow cylindrical tubes also have a thin cross section bounded by two surfaces. Lamb wave theory of plates assumes an infinite plate extent, whereas in cylinders, the circumferential curvature results in a periodic boundary condition in one dimension. This increases the complexity of Lamb waves in tubes, and many more modes of wave propagation occur in tubes than in plates. In pipes, three

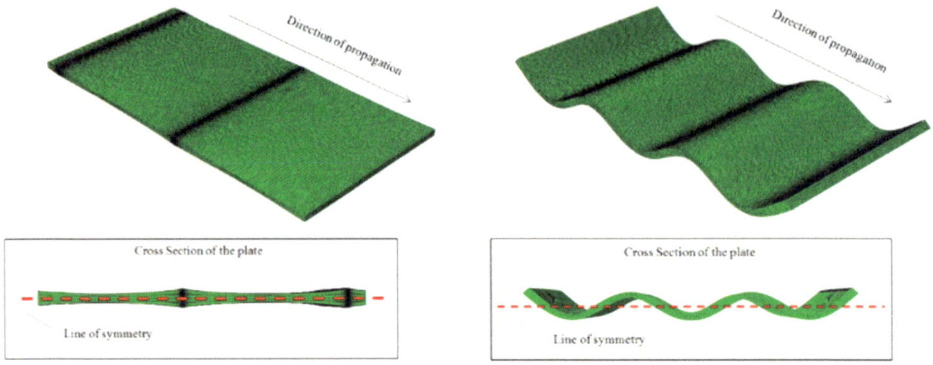

Figure 2.
Displacement of the fundamental symmetric (S0) and asymmetric (A0) wave mode. Note the displacement from the line of symmetry (red dashed line).

families of modes based on their displacement patterns are present. Axially symmetric wave modes—Longitudinal (L) and Torsional (T); and non-axially symmetric—Flexural (F) modes are illustrated in **Figure 3**. The L and T modes in cylindrical structures are analogous to the Lamb waves and SH modes of vibration in plates, respectively. The wave mode designation is defined by Meitzler [19] and includes two numbers, for example L(0,1), where the first number is the circumferential wavenumber (also known as the order) and the second number represents the sequential mode. All axially symmetric torsional and longitudinal modes are zero order modes. Flexural modes are non-axially symmetric and of order higher than zero.

Phase velocity (v_p) and group velocity (v_g) are two important terms in UGW propagation. v_p is the speed at which a continuous wave propagates. For GWT, it is important to discriminate propagating wave modes by exciting them as a discrete wave pulse with a finite number of cycles. This pulse is controlled by a window function (e.g. hamming) which comprises a bandwidth of frequencies. The speed at which this envelope of discrete pulse propagates is v_g. Variation of phase velocity with frequency leads to dispersion occurring as the UGW propagates in the structure.

At any given frequency, a number of wave modes may be present in the structure. The wave modes with frequency dependent velocities are called dispersive as they spread in space over time. Dispersion curves illustrate guided waves and their behaviour with frequency for each possible mode in the given structure. Commercial software packages [20, 21] are available to generate dispersion curves for multi-layered plates and cylindrical structures. **Figure 4** shows the dispersion curves computed for a 6 inch Schedule 40 pipe (168.3 mm outer diameter, 7.11 mm wall thickness) and a

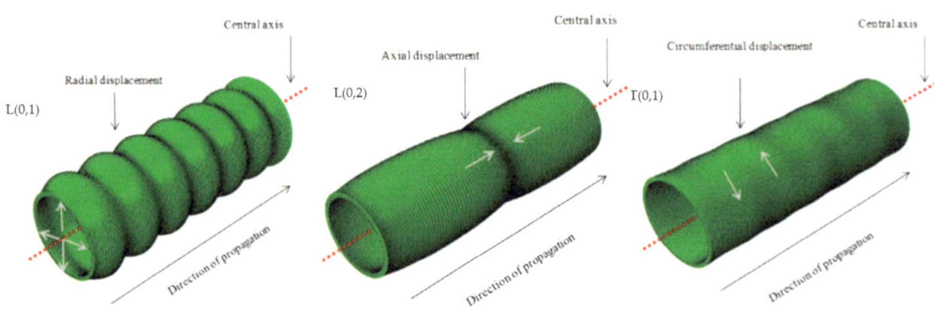

Figure 3.
Displacement of the axisymmetric L(0,1); L(0,2) and T(0,1) wave modes. Note the dominant radial, axial and circumferential displacements from the central axis, respectively.

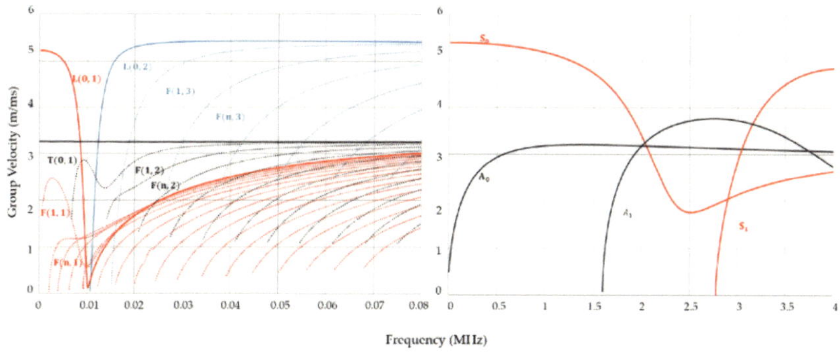

Figure 4.
Dispersion curves for 6 inch schedule 40 steel pipe (left) and 1 mm thick steel plate (right); showing the relationship between group velocity and frequency for different modes.

1 mm thick steel plate [material properties used, density (ρ)—7830 kg/m^3, Young's modulus (E)—207 GPa and Poisson's ratio (μ) = 0.3].

For the pipe, axisymmetric L(0,1), T(0,1) and L(0,2) modes are highlighted and their respective associated flexural modes, F(n,1), F(n,2) and F(n,3) are coloured red. It should be noted that L(0,2) and T(0,1) in pipes are analogous to A0 and SH0 wave modes in plates. It can be seen that the T(0,1) wave mode is completely non-dispersive for all frequencies of interest for GWT as the phase velocity dispersion curve is flat. L(0,2) is relatively non-dispersive above a certain frequency and L(0,1) is relatively dispersive in comparison to the other two axisymmetric modes. Compared to a pipe, relatively low numbers of modes are present in plates, which makes mode separation and signal interpretation much less challenging. For GWT, it is desirable to use non-dispersive wave modes for easy data interpretation.

2.2 Guided wave excitation

In contrast to conventional ultrasonic testing (UT), where high frequencies are used to examine the material directly under the test location, in GWT, low frequency ultrasound is guided through the structural boundaries and can travel tens of metres. A transducer can excite all modes that exist within its frequency bandwidth and this can complicate the received signals, making their interpretation difficult. Dispersion and the presence of multiple guided wave modes are the two main problems for GWT [22], and for practical applications, it is important for the transducer system to excite a single, non-dispersive wave mode [23]. A procedure for identifying suitable modes for a particular inspection task has been proposed by Wilcox [24] which considers the properties of the structure (dispersion, attenuation and sensitivity) and transducer (excitability, detectability and selectivity).

There are a number of different transduction technologies for excitation and detection of UGW, including Electromagnetic Acoustic Transducer (EMAT) [25], magnetostrictive devices [26], laser [27], piezoelectric and piezocomposite transducers [28]. Piezoelectric transducers offer the most promising solution due to their stability and reliability, and cost-effectiveness with simple and light-weight construction [29]. Lead zirconate titanate (PZT) has been a popular choice for UGW as it shows good electromechanical properties (electromechanical coupling, k > 0.7) which is essential to achieve large coverage. Linear and circular PZT arrays on plates have achieved inspection range of 3000 times the dimensions of the array. Application of PZT material is however limited to temperature below ~150°C (1/2 T_c) above which it experiences accelerated performance degradation over time [30]. Piezoelectric materials for SHM at higher temperatures are available [31, 32] for steamlines.

For pipes, excitation of axisymmetric wave modes [L(0,2) and T(0,1)] using piezoelectric transducers requires a circumferential ring of transducers. The circumferential spacing between the transducers in the array should be even for a high level of mode purity. All transducers in the ring are excited equally and concurrently to launch these axisymmetric modes. Apart from being non-dispersive, both of these modes provide uniform stress over the whole pipe cross section area and provide 100% coverage. Two rings of dry-coupled piezoelectric shear transducers [33] can be used to obtain unidirectional propagation of the L(0,2) mode with propagation distances approaching 50 metres. The second axisymmetric mode, L(0,1), is excited alongside L(0,2) (**Figure 5**), and can complicate the interpretation of results [34]. Therefore, an additional ring of transducers is required to suppress this undesired L(0,1) mode. This however adds to the cost of the system, significantly for larger diameter pipes. On the contrary, the T(0,1) mode is the only axisymmetric torsional mode in the frequency range of interest for GWT, so to obtain a single mode and unidirectional excitation, only two rings of transducers are required. The torsional mode requires an excitation force in the circumferential direction. This can be achieved by displacing the shear transducer used for axial longitudinal excitation by 90°. To cancel the propagation of non-axisymmetric Flexural modes, the number of transducers in a circumferential ring should be greater than the highest order of flexural mode present in the chosen frequency range [35].

For plates, the A0 Lamb mode is the easiest omnidirectional mode to excite as it only requires a point-source exerting a pure out-of-plane force on the surface of the plate. It is also the mode which has the smallest wavelength for a given frequency, therefore offering better resolution to defects compared to the S0 mode. However, due to the attenuation and higher dispersion characteristics, this mode has been predominantly neglected in favour of S0 and SH0. **Figure 6** shows the propagation of these three modes excited using uniaxial in-plane vibration.

Commercially available in-plane thickness shear transducers can generate all fundamental plate modes in the GWT operating frequency range. Both Lamb modes are generated in the axis of vibration while the SH0 mode is generated perpendicular to the axis of vibration.

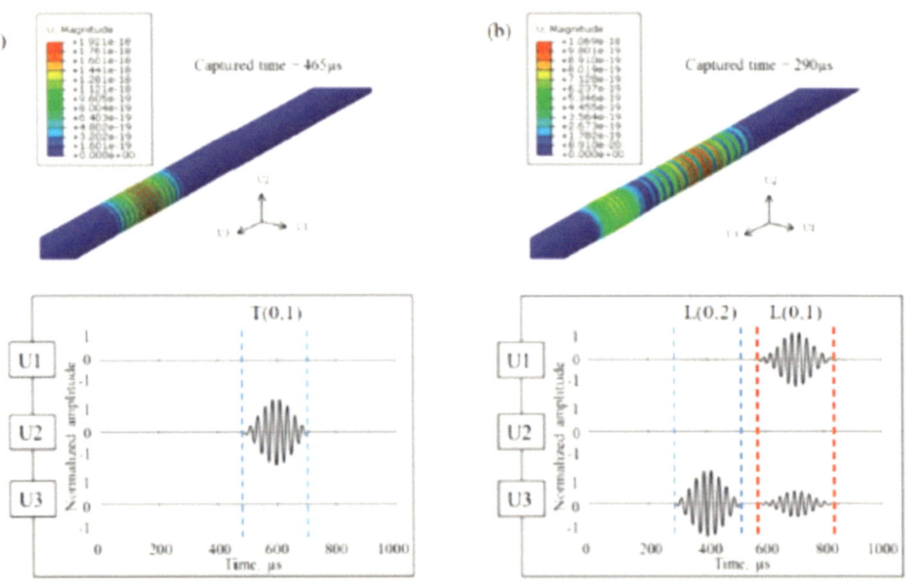

Figure 5.
Displacement patterns and waveforms generated by array of shear transducers aligned (a) circumferentially and (b) axially. U1, U2 and U3 represent radial, circumferential and axial displacement caused by transducer vibration measured using a 3D vibrometer.

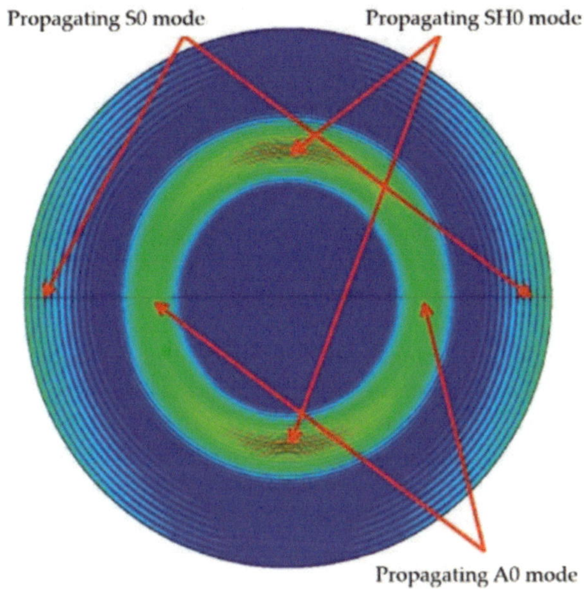

Figure 6.
Propagation of ultrasonic guided wave modes in a 3.5 m diameter, 10 mm thick steel plate from a uniaxial in-plane vibration.

2.3 Guided wave inspection

A typical GWT architecture in **Figure 7** shows the key components of the system. Apart from the transducers, the system comprises of a portable computer (PC) to control the test, and a pulser-receiver connected to the transducers to transmit and receive the ultrasonic signal to and from the structure under test. Narrow band signals such as several cycles of sine wave modulated with a window function (e.g. hamming), are generally used. These narrow band signals offer good signal strength

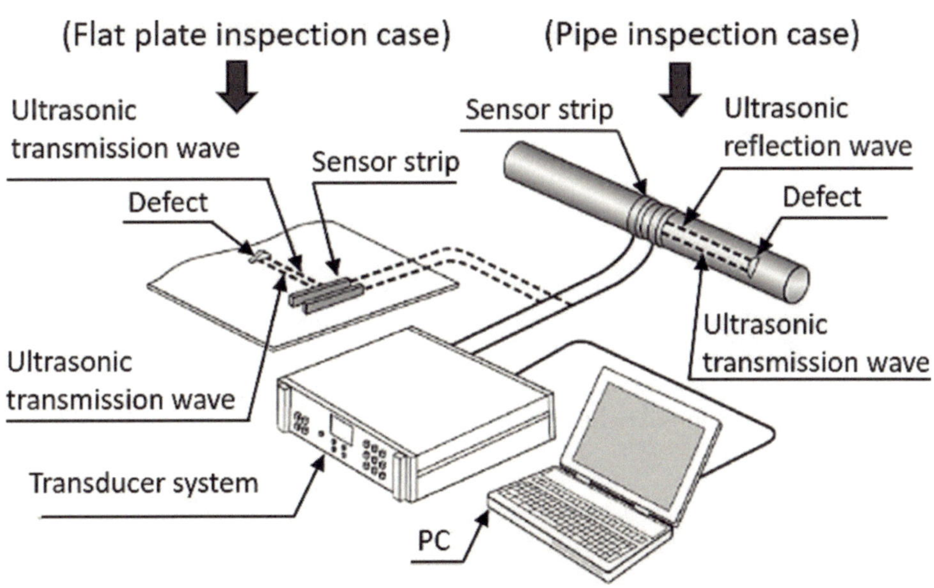

Figure 7.
Architecture of a typical guided wave inspection system [36].

and avoid dispersion while propagating long distances. The centre frequency of these signals are chosen based on the desired wave mode to achieve low dispersion over the frequencies in the narrow band.

There are two modes of operation: pulse-echo and pitch-catch. Pulse-echo mode is more common and utilises the same transducers to excite the UGW and receive the reflected signals as illustrated in **Figure 7**. Pitch-catch mode uses two sets of transducers, one to excite the UGW and the other to receive, and is only used if high resolution or a high inspection range are required. As the UGW propagates in the structure, a proportion of the energy contained in the propagating wave front will be reflected when an acoustic impedance change occurs at a feature or discontinuity in the structure. This enables full coverage of the cross section of the plate or pipe, detecting and locating both internal and external defects without disrupting operation.

Since the initial developments of GWT of pipes in late 1990s, several studies have been carried out to understand the interaction of T(0,1) and L(0,2) guided wave modes with pipe features (flanges and pipe supports) [37] and defects [35], and the effect of different defect characteristics and excitation frequencies has also been reported [38, 39]. This has led to definitions and standards for GWT instrumentation, data collection and analysis in ISO 18211:2016 [40]. When an axisymmetric mode is incident on an axisymmetric pipe feature such as a uniform weld or a flange, axisymmetric modes are reflected. With a non-axisymmetric feature such as corrosion, a non-axisymmetric wave will also be reflected back to the transducer array. The presence and axial location of defects can thus be determined by analysing these reflections and their time of arrival. Although the L(0,2) mode has shown ~2.5 times more flaw sensitivity compared to T(0,1) [34], it is difficult to excite in pure form and requires complex signal processing due to its dispersive nature. It is also affected by fluid in the pipe, so the torsional mode is more commonly used in practice. GWT using T(0,1) is most effective on straight sections achieving several tens of metres of inspection range but recent studies have evaluated its performance on bends [41].

3. Structural health monitoring

The desire to move from current periodic structural maintenance to a more cost-effective condition-based maintenance (CBM) philosophy to ensure integrity of critical structures has fostered research and development activities to develop SHM solutions. SHM using UGW has found a variety of practical applications for elongated engineering structures including pipes, plates, ship hulls, rails and cables, because of its inherent long range propagation [42].

3.1 Monitoring system design and architecture

The operational requirements of SHM systems for pipelines and storage tanks are tabulated in **Table 1**. Currently, costly acquisition of SHM data is only justifiable for structures with significantly high failure consequences. Transducer technologies play a critical role in the design of SHM system as they are permanently installed on the structure and required to repeatedly transmit excitation signals and analyse the received responses.

The transducers may need to be attached in environmentally hostile, safety-critical or difficult-to-access areas and therefore they should be designed to perform reliably under prolonged exposure to harsh environmental and operational conditions (EOCs). Therefore, low cost and reliability are the two main factors to consider when designing a SHM sensor system for pipelines and storage tanks. One

Operational requirement	Pipelines	Storage tank floor
Operating temperature	−10 to 150°C	−10 to 60°C
Signal to noise	<6 dB	<6 dB
Operating frequency range	20–100 kHz	Resonant frequency
Transmission range	Up to 100 m	30–100 m
Frequency of data collection	Once a week	Once a week (depending on the condition)
Wave mode selection	T(0,1)	S0 and SH0
Signal processing	Thresholding/outlier analysis Baseline subtraction Pattern recognition	Tomography Pattern recognition Neural networking Baseline subtraction
Data acquisition	Pulse-echo/pitch-catch	Pitch-catch

Table 1.
Operational requirements of SHM systems.

cost-effective approach is to use a single pulser-receiver and PC to collect monitoring data from multiple sensor locations at junction points, which can be located in easily accessible location. This significantly reduces cost of repeated access and of the overall system.

3.2 Monitoring system for pipelines

Current state-of-the-art in pipeline monitoring solutions includes corrosion coupons, acoustic emission and magnetostrictive sensors, flexible eddy current arrays, flexible ultrasonic transducers, guided wave sensors, impedance spectroscopy, microwave backscattering and fibre optic sensors. A review of these monitoring technologies can be found in [43]. Corrosion sensors based on electrical resistance and electromechanical impedance spectroscopy can only provide coverage over a small area and are not suitable for non-uniform corrosion artefacts such as pits. Recent advances in acoustic emission (AE) sensor technology [44, 45] have led to corrosion detection and monitoring solutions where acoustics signals from micro-fractures and delamination of the oxide are analysed. These emissions release much less energy than emission from crack growth where AE has shown great potential. In low noise environments AE could be used to detect signals from corrosion with tens of metres range using monitoring frequencies of tens of kilohertz. However, in a live plant, high process noise requires several hundred kilohertz of monitoring frequencies and coverage is limited <0.5 m and requires complex signal processing. For this reason, AE is limited for this application. Magnetostrictive sensor (MsS) is another technology for pipeline monitoring first developed and patented by SwRI® [46]. They have lower power output compared to piezoelectric transducer, however, recent advancements have reported significant improvements in their power output, sensitivity and flaw characterisation [47]. Piezoelectric sensing offers the most promising solution due to their stability, reliability, and cost-effectiveness as described in Section 2.2. This has enabled the development of several SHM solutions. Guided Ultrasonic Ltd. offers one such monitoring system gPIMS [48] and this system's stability and defect detection capabilities have been demonstrated [49] at temperatures up to 90°C. Another example is the system developed by the authors and its installation, operation and performance is reported [50]. **Figure 8** shows some of these pipeline monitoring devices.

3.3 Monitoring system for storage tanks

Monitoring of a tank floor is more important compared to the tank wall, due to the fact that degradation of the tank floor is not visible until it becomes severe. A tank floor comprises a large number of plates (dependent on the tank diameter) of 6–8 mm thickness joined with lap welds. SHM of tank floors using UGW is challenging due to this complicated layout, the propagation distance requirement, level of attenuation, and wave reflections and mode conversions at boundaries. GWT of above-ground storage tanks (AST) is an emerging technology and was first explored in 2006 by Mažeika et al. [17]. S0 mode was chosen as the principal mode of interest due to low energy losses from the fluid inside the tank compared to A0 mode [51]. Considering the large area and complexity of tank floor designs, guided waves should be transmitted with as much energy as possible. To achieve full coverage, a Pitch-Catch configuration (through transmission) is preferred for data acquisition and the appropriate transducer array layout was studied by Mažeika et al. [17] and Feng et al. [52].

Transducer bonding is also problematic as the tank annular chime gets heavily corroded over time due to environmental influences. Previous studies on selection of sensor location have evaluated two scenarios: wave excitation on tank annular chime; and tank wall. Currently, normal mode transducers (elongated type) are installed on the annular chime of the tank to transmit guided waves across the floor plate, and a tomographic technique is used to map the structural health of the tank floor [53]. The SH0 mode is an interesting alternative to the S0 mode for this application due to its non-dispersive characteristic [54]. Advances in flexible shear mode transducers led to a recent study [55] on their application to SHM of AST floors. This study evaluated the two modes of interest: S0 mode for normal excitation; and SH0 mode for shear excitations. Sensor location on both the tank wall and annular chime were considered for the two modes. The sensor location is illustrated in **Figure 9**.

The wave propagation for both cases is illustrated in **Figure 10**. A significant amplitude drop for the applied normal load on the tank wall was observed in comparison to the tank floor. However, in the case of shear loading, insignificant amplitude drop was observed.

The application of shear stress on tank wall for guided wave testing of tank floors was thus realised. This increases the potential market for tank floor inspection using UGW as the tank wall can be used to bond shear transducers for structural health monitoring.

Figure 8.
Commercial pipeline SHM systems (left to right): MsS [46], gPIMS [48] and iPerm [50].

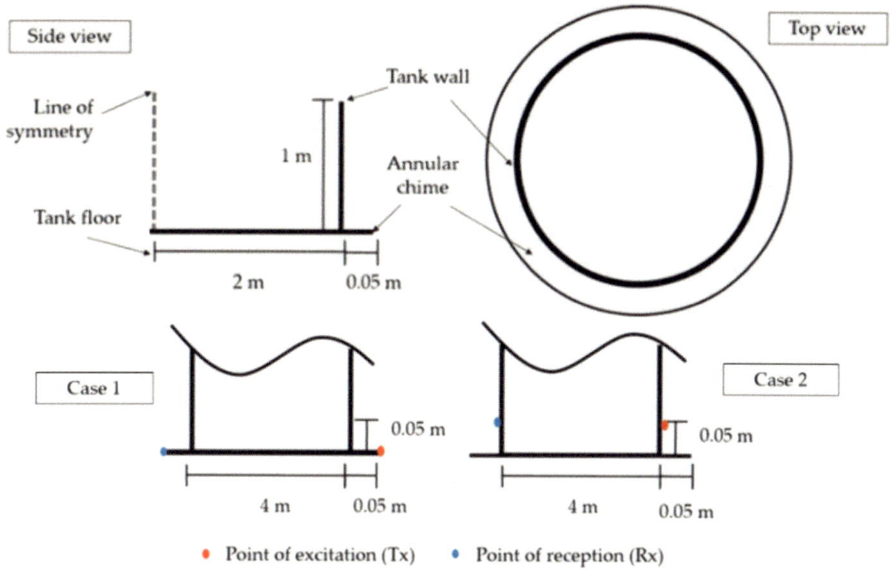

Figure 9.
Schematic of a tank (top) and layout of the point of excitation and reception of the two cases studied (bottom)—excitation and reception from the tank floor in Case 1 and tank wall in Case 2.

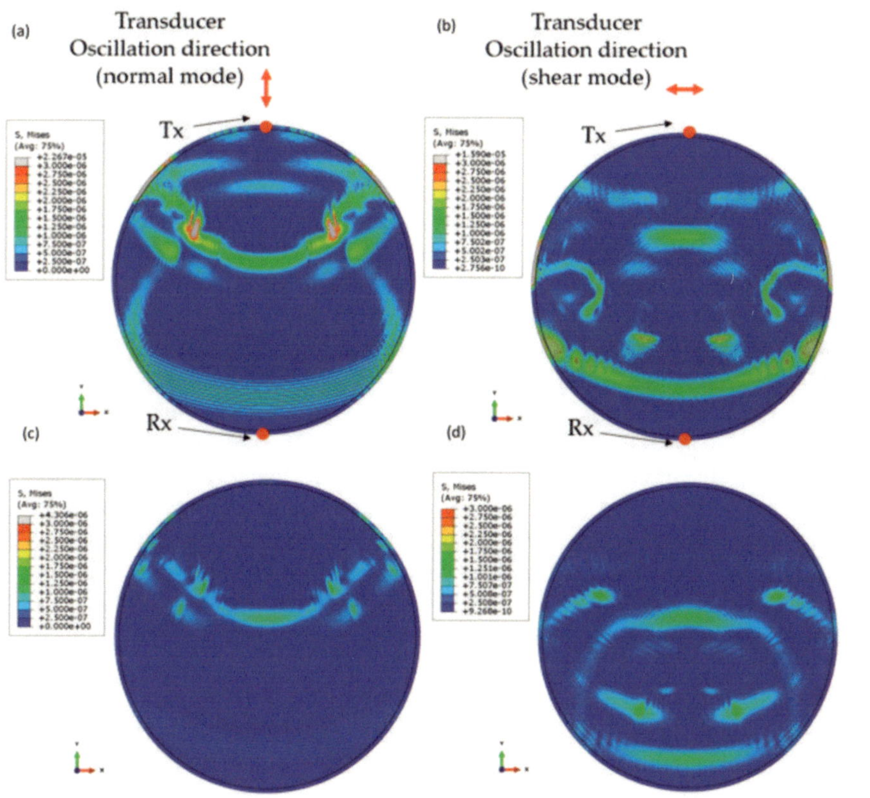

Figure 10.
FEA showing UGW excitation on tank annular chime and tank wall: applied (a) normal and (b) shear stress on tank chime; and (c) normal and (d) shear stress on tank wall.

4. Data processing for SHM

4.1 Effect of environmental and operating conditions

Several investigations into the effect of environmental and operational conditions on the recorded ultrasonic signals have been carried out, and change of temperature has been shown to be the main source of signal fluctuations [56–58]. The influence of temperature on GWT is a combination of effects due to the structure's mechanical properties and the effects on ultrasonic transducers and their bonding. Previous study has reported that, for small ambient temperature variations of a few degrees, the effect on transducer performance is much less significant than that on the wave propagation [59]. The UGW signals will undergo changes in the amplitude and phase. The change in UGW signal amplitude is attributable to changes in temperature-dependent properties of the ultrasonic transducer, particularly the piezoelectric materials and adhesives. To minimise this variability, careful selection of adhesives and transducer material for target temperature is recommended. The phase shift in the UGW signal is due to the change in wave propagation velocity due to variation in the mechanical properties of the waveguide [60], i.e. pipe or tank floor in this study. The material properties of relevance include elastic and shear moduli, and the density; which in turn relates to the elasto-acoustic properties of the material, acoustic absorption and ultrasonic wave velocity. Thermal expansion adds to this effect by changing the propagation distance directly and indirectly through changes in the thickness of the plate or the pipe. The relationship between the difference in time of arrival (TOA) of the signal and the change in temperature of the structure can be described as:

$$\delta t = \frac{d}{v}(\alpha - \gamma)\delta \tag{1}$$

where δt is the difference in TOA of the signals when the change in temperature of the structure is δT. d is the distance travelled by the wave and v is the wave velocity. α is the coefficient of thermal expansion and γ is the coefficient of change in phase velocity. γ is generally much greater than α and hence from Eq. (1), it can be seen that the main contribution to change of TOA is from change in wave velocity due to temperature variations. Also, since the time shift is directly proportional to the propagation distance, it can be noted that the effect of temperature on UGW will increase with propagation distance. This can be significant for the large propagation distances required for pipes. The inverse relationship between temperature and wave velocity suggests that faster modes will be less affected than slower ones. These temperature induced variations in UGW signals can adversely reduce the defect detection capabilities of the SHM system. An experimental study [61] showed that the effect of temperature variation on UGW from ambient temperature up to 70°C was much more pronounced than the effect of a drilled hole of 1 mm diameter.

4.2 Temperature compensation algorithms

The issues described in Section 4.1 led to several investigations within the SHM research community and a number of EOC compensation strategies have been proposed. Their main objective is to achieve UGW propagation time and amplitude correction for enhanced defect sensitivity. These correction strategies can be classified into two techniques: data-driven and analytical physics-based.

The data-driven techniques requires a large set of baseline measurements from the structure at different temperatures. A signal from the 'bank' of baselines is then selected to minimise the difference relative to the test signal for a particular temperature. This method is called Optimum Baseline Selection (OBS). A number of selection criteria including mean square deviation [56] and maximum residual amplitude [62] have been proposed. This method has limitations for cases when a large set of baselines is not available and if the temperature of the selected baseline is different from the temperature of the test signal. Baseline signal stretch (BSS) was introduced as a complimentary technique that in its simplest form requires only one single baseline at a reference temperature. In BSS, time domain stretching is performed to adjust the selected baseline and the local coherence is estimated as a function of time. BSS can be performed in both time and frequency domain to achieve similar performance [63]. A number of researchers have explored these methods to provide enhanced temperature compensation with a reduced number of baseline data sets [62–65]. The temperature resolution of the baseline set is defined by the capability of BSS method as the stretching required for large temperature difference leads to distortion of the signal's frequency content. The performance of BSS depends on signal complexity and mode purity and. For practical application, a temperature step of 1–2°C is recommended for baseline dataset [63]. Recently developed modified-BSS (MBBS) method outperformed BSS and is more effective for temperature differences of up to 13°C [66]. BSS can be computation intensive and alternative methods with improved computational speed have been proposed that operate on signals in the stretch factor and scale-transform domain [67].

Physics-based analytical techniques for temperature compensation [68, 69] utilise underlying physical principles such as changes in material properties and thermal expansion (described in Section 4.2) for transducer signal reconstruction at different temperatures. The advantage of these techniques is that it does not require a large set of baseline sensor measurements from the structure. The performance of these analytical temperature compensation models is shown to be at par with the state-of-the-art data driven techniques. They are however limited to simple structural geometries and boundary conditions. Combinations of analytical and data-driven strategies that require fewer baselines are being explored [70] which will offer an efficient, practical and useful approach for temperature compensation.

4.3 Damage detection strategies

A method for damage detection must be applied to the corrected data to see whether the structure being monitored has developed any damage. In structures containing high densities of structural elements, the time-traces obtained are often too complex to be directly interpreted due to a large number of overlapping reflections. A popular approach for SHM is *baseline subtraction*, which is based on the comparison of structure's ultrasonic response at original state (baseline) with response at a later stage. The subtracted *residual signal* will remove reflections from pipe or tank floor features and isolate any damage scattered signals as illustrated in **Figure 11**.

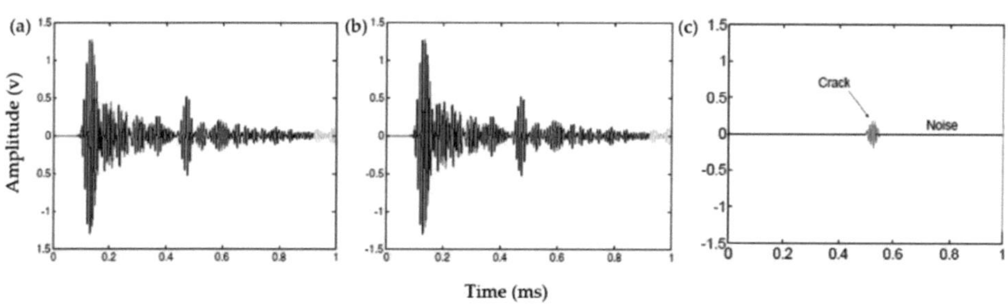

Figure 11.
Baseline subtraction of UGW time traces (a) undamaged structure (b) damaged structure (c) defect signal after baseline subtraction [71].

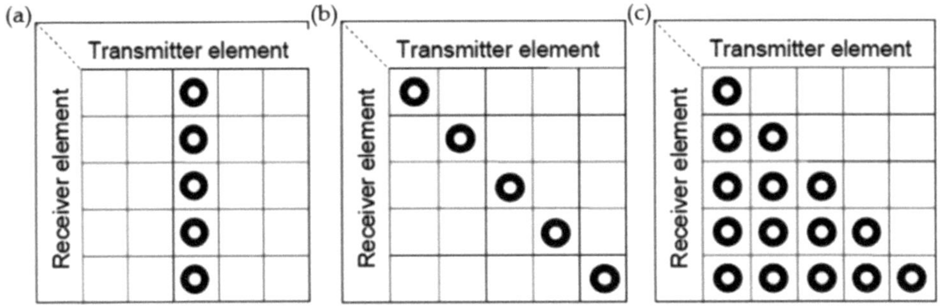

Figure 12.
Transmit-receive matrices for the imaging algorithms; (a) common source method (b) synthetic aperture focusing technique and (c) total focusing method [72].

For sensor arrays Full-Matrix Capture (FMC) is a data acquisition process which records all possible transmit-receive combinations of UGW data. This data collection matrix is symmetric due to reciprocity (**Figure 12**) and only the lower and upper triangular parts of the matrix need be recorded. This data can then be used to obtain tomography images of the structure or perform sound energy focusing techniques to improve SNR.

For complex structures and if the data corresponding to the damage state is not known *a priori*, damage detection strategies based on unsupervised algorithms are used. One such strategy is based on the Outlier Analysis (OA) algorithm which extracts damage sensitive features from the UGW signals and aims to identify if they have deviated from their baseline distribution using Mahalanobis squared distance [73]. OA can be applied as univariate and multivariate depending on a number of features. For univariate implementation, root mean square (RMS) of the signal has been successfully used as a damage sensitive feature for detection of corrosion type defects in plates [56] and pipes [74]. To increase the damage sensitivity, multivariate OA is recommended, where a number of features are extracted from the UGW signals and classical methods of multivariate statistics such as principal component analysis (PCA) are applied. For UGW, the features of interest include time-of-flight, frequency centres, energies, modes of scattered waves, and time-frequency spread. A review of the feature extraction approaches based on time-frequency representations such as short-time Fourier transform, Wigner-Ville distribution, Hilbert-Huang transform, and wavelet transform can be found in [75]. Recent advances in the field of artificial intelligence led to researchers formulating defect detection as a machine learning problem. A study using an Artificial Neural Network (ANN) based strategy was applied for damage classification [73] and was reported to outperform OA for damage detection using just one feature. Such supervised machine learning strategies will however require data from the structure with known types and levels of damage, which may not always be present.

5. Conclusions

This chapter presents the advances in guided wave technology for structural health monitoring of two of the most critical metallic assets, pipelines and storage tanks, in the Oil and Gas industry. These SHM technologies support cost-effective asset integrity management by enabling a condition based maintenance model, moving away from conventional routine inspection. The advances in SHM technologies of pipes and tanks are presented. Operational requirements of these SHM systems are discussed with a thorough review of the state-of-the-art and fundamentals of pipelines and tank floor inspection using UGW. Limitations of SHM for high temperature pipelines have also been identified for future research and development.

Author details

Anurag Dhutti[1], Shehan Lowe[1] and Tat-Hean Gan[1,2*]

1 Brunel University London, Uxbridge, Middlesex, UK

2 TWI Ltd, Cambridge, UK

*Address all correspondence to: tat-hean.gan@brunel.ac.uk

References

[1] Wood MH, Arellano ALV, Van Wijk L. Corrosion related accidents in petroleum refineries. Eur Comm Jt Res Centre, Rep no EUR. 2013:26331. DOI: 10.2788/37909

[2] China oil pipe blast: Qingdao pipeline blast "kills 44" – BBC News [Internet]. Available from: https://www.bbc.co.uk/news/world-asia-china-25050300 [Accessed: October 31, 2018]

[3] Chang JI, Lin C-C. A study of storage tank accidents. Journal of Loss Prevention in the Process Industries. 2006;**19**:51-59. DOI: 10.1016/j.jlp.2005.05.015

[4] Fire rages for fourth day at fuel storage facility in Brazil's largest port Santos – Society's Child – Sott.net [Internet]. Available from: https://www.sott.net/article/294831-Fire-rages-for-fourth-day-at-fuel-storage-facility-in-Brazils-largest-port-Santos [Accessed: October 31, 2018]

[5] Anderson MT, Kunerth DC, Davidson JR. Nondestructive examination technologies for inspection of radioactive waste storage tanks. 1995;**27**(8). DOI: 10.2172/114637. Available from: https://inis.iaea.org/search/search.aspx?orig_q=RN:27026430

[6] Su Z, Ye L. Identification of Damage Using Lamb Waves: From Fundamentals to Applications. London: Springer; 2009. p. 346. DOI: 10.1007/978-1-84882-784-4

[7] Gazis DC, Herman R, Wallis RF. Surface elastic waves in cubic crystals. Physics Review. 1960;**119**:533-544. DOI: 10.1103/PhysRev.119.533

[8] Zemanek J. An experimental and theoretical investigation of elastic wave propagation in a cylinder. The Journal of the Acoustical Society of America. 1972;**51**:265-283. DOI: 10.1121/1.1912838

[9] Silk MG, Bainton KF. The propagation in metal tubing of ultrasonic wave modes equivalent to Lamb waves. Ultrasonics. 1979;**17**:11-19. DOI: 10.1016/0041-624X(79)90006-4

[10] Alleyne D, Cawley P. The practical excitation and measurement of Lamb waves using piezoelectric transducers. Review of Progress in Quantitative Nondestructive Evaluation. 1994;**13**:181

[11] Mudge PJ, Lank AM, Alleyne DN. A long range method of detection of corrosion under insulation in process pipework. Journal of Japanese Society for Non-Destructive Inspection (JSNDI). 1997;**46**:314-319

[12] Alleyne DN, Cawley P, Lank AM, Mudge PJ. The Lamb wave inspection of chemical plant pipework. In: Review of Progress in Quantitative Nondestructive Evaluation. Boston, MA: Springer US; 1997. pp. 1269-1276. DOI: 10.1007/978-1-4615-5947-4_165

[13] Eddyfi. Long-Range Guided Wave Testing with Teletest Focus+ [Internet]. 2018. Available from: https://www.teletestndt.com/ [Accessed: October 06, 2018]

[14] Guilded Ultrasonic Ltd. Inspection [Internet]. Available from: http://www.guided-ultrasonics.com/inspection/ [Accessed: October 06, 2018]

[15] Worlton DC. Ultrasonic testing with Lamb waves. Technical Report. GE US. 1956

[16] Viktorov IA. Rayleigh and Lamb Waves. Boston, MA: Springer US; 1967. p. 154. DOI: 10.1007/978-1-4899-5681-1

[17] Mažeika L, Kažys R, Raišutis R, Demčenko A, Šliteris R. Long-range ultrasonic non-destructive testing of fuel tanks. In: Proceedings of 9th ECNDT. Berlin; 25-29 Sept 2006. pp. 1-8

[18] Lamb H. On waves in an elastic plate. Proceedings of the Royal Society A: Mathematical, Physical and Engineering Science. 1917;**93**: 114-128. DOI: 10.1098/rspa.1917.0008

[19] Meitzler AH. Mode coupling occurring in the propagation of elastic pulses in wires. The Journal of the Acoustical Society of America. 1961;**33**:435-445. DOI: 10.1121/1.1908685

[20] Lowe MJS. A model for predicting the properties of guided ultrasonic waves, and illustration of its application to NDT. In: IEE Colloquium New Applications in Modelling and Inversion Techniques for NDT; 29 Jan 1999; London. DOI: 10.1049/ic:19990111

[21] Bocchini P, Marzani A, Viola E. Graphical user Interface for guided acoustic waves. Journal of Computing in Civil Engineering. 2011;**25**:202-210. DOI: 10.1061/(ASCE) CP.1943-5487.0000081

[22] Wilcox PD, Lowe MJS, Cawley P. The effect of dispersion on long-range inspection using guided waves. NDT and E International. 2001;**34**:1-9. DOI: 10.1016/S0963-8695(00)00024-4

[23] Galvagni A, Cawley P. The reflection of guided waves from simple support in pipes. The Journal of the Acoustical Society of America. 2011;**129**:1869-1880. DOI: 10.1121/1.3543958

[24] Wilcox PD, Lowe MJS, Cawley P. Mode & transducer selection for long range Lamb wave inspection. Journal of Intelligent Material Systems and Structures. 2001;**12**:553-565. DOI: 10.1106/N9PB-Y62E-P0Y2-50QF

[25] Clough M, Fleming M, Dixon S. Circumferential guided wave EMAT system for pipeline screening using shear horizontal ultrasound. NDT&E international. 2017;**86**:20-27. DOI: 10.1016/j.ndteint.2016.11.010

[26] Zhu L, Wang Y, Sun F. Single torsional guided wave excitation in pipes by frequency selection using magnetostrictive sensor technology. ICEMI 2013. 2013;**2**:872-876. DOI: 10.1109/ICEMI.2013.6743158

[27] Jeon JY, Jung HK, Kim DH, Park G. Experimental Study on Guided & Standing Waves Based Full Field Laser Scanning for Damage Detection. 9th EWSHM; 10-13 Jul 2018; Manchester, UK

[28] Gao H. Ultrasonic guided wave annular array transducers for structural health monitoring. AIP Conference Proceedings. 2006;**25**:1680-1686. DOI: 10.1063/1.2184723

[29] Jiang X, Kim K, Zhang S, Johnson J, Salazar G. High-temperature piezoelectric sensing. Sensors (Switzerland). 2013;**14**:144-169. DOI: 10.3390/s140100144

[30] Gotmare SW, Leontsev SO, Eitel RE. Thermal degradation and aging of high-temperature piezoelectric ceramics. Journal of the American Ceramic Society. 2010;**93**:1965-1969. DOI: 10.1111/j.1551-2916.2010.03663.x

[31] Stevenson T, Martin DG, Cowin PI, Blumfield A, Bell AJ, Comyn TP, et al. Piezoelectric materials for high temperature transducers and actuators. Journal of Materials Science: Materials in Electronics. 2015;**26**:9256-9267. DOI: 10.1007/s10854-015-3629-4

[32] Zhang S, Yu F. Piezoelectric materials for high temperature sensors. Journal of the American Ceramic Society. 2011;**94**:3153-3170. DOI: 10.1111/j.1551-2916.2011.04792.x

[33] Alleyne DN, Cawley P. The excitation of Lamb waves in pipes using dry-coupled piezoelectric transducers. Journal of Nondestructive Evaluation. 1996;**15**:11-20. DOI: 10.1007/BF00733822

[34] Lowe PS, Sanderson RM, Boulgouris NV, Haig AG, Balachandran W. Inspection of cylindrical structures using the first longitudinal guided wave mode in isolation for higher flaw sensitivity. IEEE Sensors Journal. 2016;**16**:706-714. DOI: 10.1109/JSEN.2015.2487602

[35] Lowe MJSJS, Alleyne DNN, Cawley P. Defect detection in pipes using guided waves. Ultrasonics. 1998;**36**:147-154. DOI: 10.1016/S0041-624X(97)00038-3

[36] Kuroishi T, Sakata F. New application of non-destructive inspection technology guided wave pipe inspection and monitoring system. Journal of the Japan Welding Society. 2006;**75**:220-224. DOI: 10.2207/jjws.75.220

[37] Alleyne DN, Cawley P. The effect of discontinuities on the long-range propagation of Lamb waves in pipes. Proceedings of the Institution of Mechanical Engineers - Part E: Journal ofProcess Mechanical Engineering. 1996;**210**:217-226. DOI: 10.1243/PIME_PROC_1996_210_316_02

[38] Cawley P, Lowe MJS, Simonetti F, Chevalier C, Roosenbrand AG. The variation of the reflection coefficient of extensional guided waves in pipes from defects as a function of defect depth, axial extent, circumferential extent and frequency. Proceedings of the Institution of Mechanical Engineers, Part C: Journal of Mechanical Engineering Science. 2002:1131-1143. DOI: 10.1243/095440602761609498

[39] Rojas E, Baltazar A, Treesatayapun C. Investigation on damage identification in a pipe using torsional guided waves. In: AIP Conference Proceedings. 2017. DOI: 10.1063/1.4974586

[40] ISO 18211:2016. Non-destructive testing–Long-range inspection of above-ground pipelines and plant piping using guided wave testing with axial propagation; 2016

[41] Heinlein S, Cawley P, Vogt TK. Reflection of torsional T(0,1) guided waves from defects in pipe bends. NDT&E international. 2018;**93**:57-63. DOI: 10.1016/j.ndteint.2017.09.007

[42] Rose JL. Ultrasonic guided waves in structural health monitoring. Key Engineering Materials. 2004; **270-273**:14-21. DOI: 10.4028/www.scientific.net/KEM.270-273.14

[43] Liu Z, Kleiner Y. State-of-the-art review of technologies for pipe structural health monitoring. IEEE Sensors Journal. 2012;**12**:1987-1992. DOI: 10.1109/JSEN.2011.2181161

[44] Tscheliesnig P, Lackner G, Jagenbrein A. Corrosion detection by means of acoustic emission monitoring. In: 19th WCNDT; 13-17 June 2016; Munich Germany

[45] Cole PT, Watson JR. Acoustic emission for corrosion detection. Advances in Materials Research. 2006;**13-14**:231-236. DOI: 10.4028/www.scientific.net/AMR.13-14.231

[46] Kwun H, Holt AE. Feasibility of under-lagging corrosion detection in steel pipe using the magnetostrictive sensor technique. NDT&E international. 1995;**28**:211-214. DOI: 10.1016/0963-8695(95)00019-T

[47] Vinogradov S, Leonard J. Development of Magnetostrictive sensor Technology for Guided Wave Examinations of piping. In: 10th ECNDT, 7-11 June 2010; Moscow, Russia

[48] Guided Ultrasonics Ltd. gPIMS [Internet]. Available from: https://www.guided-ultrasonics.com/gpims/ [Accessed: October 05, 2018]

[49] Cawley P, Cegla F, Galvagni A. Guided waves for NDT and permanently-installed monitoring. Insight – Non-Destructive Testing and Condition Monitoring.

2012;**54**:594-601. DOI: 10.1784/insi.2012.54.11.594

[50] Dhutti A, Gan TH, Kanfoud J. iPerm – A structural health monitoring device for pipelines In: 9th EWSHM; 10-13 July 2018; Manchester, UK

[51] Raišutis R, Kažys R, Mažeika L, Šliteris R. Application of the ultrasonic transmission tomography for inspection of the petroleum tank floor. Ultragarsas. 2007;**62**:26-32

[52] Feng Z, Wang W, Tong W, Yuan K, Han Z, Chen Y. Storage tank floor and wall defect in-situ inspection with ultrasonic guided wave technique. In: 8th International Pipeline Conference. 1 Oct 2010. pp. 229-232. DOI: 10.1115/IPC2010-31065

[53] Mažeika L, Kažys R, Raišutis R, Šliteris R. Guided wave tomography for inspection of the fuel tanks floor. International Journal of Materials and Product Technology. 2011;**41**:128-139. DOI: 10.1504/IJMPT.2011.040291

[54] Rose JL. Standing on the shoulders of giants: An example of guided wave inspection. Materials Evaluation. 2002;**60**:53-59

[55] Lowe PS, Duan W, Kanfoud J, Gan T-H. Structural health monitoring of above-ground storage tank floors by ultrasonic guided wave excitation on the Tank Wall. Sensors 2017;**17**:2542. DOI: 10.3390/s17112542

[56] Lu Y, Michaels JE. A methodology for structural health monitoring with diffuse ultrasonic waves in the presence of temperature variations. Ultrasonics. 2005;**43**:717-731. DOI: 10.1016/j.ultras.2005.05.001

[57] Mazzeranghi A, Vangi D. Methodology for minimizing effects of temperature in monitoring with the acousto-ultrasonic technique. Experimental Mechanics. 1999;**39**:86-91. DOI: 10.1007/BF02331110

[58] Konstantinidis G, Drinkwater BW, Wilcox PD. The temperature stability of guided wave structural health monitoring systems. Smart Materials and Structures. 2006;**15**:967-976. DOI: 10.1088/0964-1726/15/4/010

[59] Konstantinidis G, Wilcox PD, Drinkwater BW. An investigation into the temperature stability of a guided wave structural health monitoring system using permanently attached sensors. IEEE Sensors Journal. 2007;**7**:905-912. DOI: 10.1109/JSEN.2007.894908

[60] Wilcox PD, Konstantinidis G, Croxford AJ, Drinkwater BW. Strategies for guided wave structural health monitoring. AIP Conference Proceedings. 2007;**894**:1469-1476. DOI: 10.1063/1.2718139

[61] Lee BC, Staszewski WJ. Modelling of Lamb waves for damage detection in metallic structures: Part II. Wave interactions with damage. Smart Materials and Structures. 2003;**12**:815-824. DOI: 10.1088/0964-1726/12/5/019

[62] Clarke T, Simonetti F, Cawley P. Guided wave health monitoring of complex structures by sparse array systems: Influence of temperature changes on performance. Journal of Sound and Vibration. 2009;**329**:2306-2322. DOI: 10.1016/j.jsv.2009.01.052

[63] Croxford AJ, Moll J, Wilcox PD, Michaels JE. Efficient temperature compensation strategies for guided wave structural health monitoring. Ultrasonics. 2010;**50**:517-528. DOI: 10.1016/j.ultras.2009.11.002

[64] Croxford AJ, Wilcox PD, Drinkwater BW, Konstantinidis G. Strategies for guided-wave structural health monitoring. Proceedings of the Royal Society A: Mathematical,

Physical and Engineering Science. 2007;**463**:2961-2981. DOI: 10.1098/rspa.2007.0048

[65] Wilcox PD, Croxford AJ, Michaels JE, Lu Y, Drinkwater BW. A comparison of temperature compensation methods for guided wave structural health monitoring. AIP Conference Proceedings. 2008;**975**:1453-1460. DOI: 10.1063/1.2902606

[66] Salmanpour M, Sharif Khodaei Z, Aliabadi M. Guided wave temperature correction methods in structural health monitoring. Journal of Intelligent Material Systems and Structures. 2016: 1-15. DOI: 10.1177/1045389X16651155

[67] Harley JB, Moura JMF, Fry FEJ. Scale transform signal processing for optimal ultrasonic temperature compensation. IEEE Transactions on Ultrasonics, Ferroelectrics, and Frequency Control. 2012;**59**:2226-2236. DOI: 10.1146/annurev.ph.20.030158.001231

[68] Dan CA, Kudela P, Ostachowicz W, et al. Compensation of temperature effects on guided wave based structural health monitoring systems. In: 7th EWSHM; 8-11 July 2014; Nantes, France

[69] Roy S, Lonkar K, Janapati V, Chang F-K. A novel physics-based temperature compensation model for structural health monitoring using ultrasonic guided waves. Structural Health Monitoring. 2014;**13**:321-342. DOI: 10.1177/1475921714522846

[70] Neerukatti RK, Hensberry K, Kovvali N, Chattopadhyay A. A novel probabilistic approach for damage localization and prognosis including temperature compensation. Journal of Intelligent Material Systems and Structures. 2016;**27**:592-607. DOI: 10.1177/1045389X15575084

[71] Clarke T. Guided Wave Health Monitoring of Complex Structures. Department of Mechanical Engineering, Imperial College London; 2009

[72] Davies JO. Inspection of pipes using low frequency focused guided waves. [Thesis]. Imperial College London; 2008

[73] Bagheri A, Pistone E, Rizzo P. Outlier analysis and artificial neural network for the noncontact nondestructive evaluation of immersed plates. Research in Nondestructive Evaluation. 2015;**26**:154-173. DOI: 10.1080/09349847.2015.1022677

[74] Mountassir MEL, Yaacoubi S, Mourot G, Maquin D. Studies on data correction of structural health monitoring using ultrasonic guided waves: Case of study. In: Proceedings 8th EWSHM Conference; 5-8 July 2016; Bilbao, Spain

[75] Raghavan A. Guided-wave structural health monitoring. [Thesis]. University of Michigan; 2007

Chapter 6

Application of a Frequency-Based Detection Method for Evaluating Damaged Concrete Sleepers

Kodai Matsuoka and Tsutomu Watanabe

Abstract

Frequency-based damage detection (FDD) has been studied for a long time. Generally, it is pointed out that FDD is less sensitive to detect the damage in civil structures, which are composed of many members precisely. However, for the structural members on the premise of replacement like concrete sleepers, the FDD approach that has been accumulated so far may be effective. In addition, its ease and simplicity of the system are an advantage of realizing regularly and inexpensive inspection on the sites. Here we introduce the damage influence on the concrete sleepers based on the laboratory tests and demonstration of the practical use of FDD through some filed tests.

Keywords: concrete sleeper, frequency, damage detection, field test

1. Introduction

A railway prestressed concrete sleeper is one of the important components supporting rails and distributing train axle loads into ballast [1]. Mono-block type concrete sleepers, one of the most common sleeper types, are conventionally used in countries such as Australia, Canada, China, Italy, the UK, the USA, and Japan [2]. In Japan, a large number of these sleepers have been continuously introduced into railway lines since the 1950s [3], with some exceeding 50 years of operational lifetime. Thus, sound methodologies and tools for achieving effective maintenance of such a vast number of concrete sleepers are ultimately required [4].

An impact axle load is a typical scenario that can promote cracks and influence the deterioration of concrete sleepers [5–9]. In particular, some impact loads due to wheel irregularities (e.g., flat wheels), rail irregularities, and rail joints would likely over time produce concrete cracks at the bottom of rail seats and at the top of midspans where the maximum or minimum bending moment arises as shown in **Figure 1** [3, 7]. A low-occurrence probability of such significant loads acting upon concrete sleepers actually exists; however, when they do occur, yielding of steel members, residual displacement, and/or open cracking of surrounding concrete will often result. Open cracking allows water to penetrate into the sleeper ultimately causing corrosion of reinforced steel, which then leads to a loss in bending strength [8].

Currently in Japan, the inspection of concrete sleeper deterioration/damage is typically carried out by visual inspection via foot patrol. However, concrete sleeper bodies, with the exclusion of their top surfaces, are usually covered by ballast (as shown in **Figure 1**). It is therefore difficult, in general, to detect damage by visual inspection.

Even if the ballast around sleepers is scraped out, damage such as cracks cannot always be detected due to the clogging of cracks with soil dust. Moreover, scraping out ballast to inspect sleeper states is not realistic due to the enormous number of concrete sleepers that require inspection. Vibration-based structural damage detection, however, is a potential method that may be employed for effectively remedying this challenge.

Vibration-based structural damage detection is a well-known concept and widely invoked within the domain of structural health monitoring [10, 11]. For civil engineered structures, many researchers and engineers have performed such related assessments [12, 13]. Thus far, varying degrees of success for state evaluations of actual structures focusing on local and higher vibration modes and vibration characteristic changes before and after earthquakes have been reported [14–16]. In contrast, however, substantial numbers of structural damage occurrences have been detected impractically, thereby resulting in general detection methodologies that do not specifically focus on actual structural systems or accrued associated damage thereof. Assuming a damage detection method for practical applications, however, characteristics of target structure types and their typically incurred damage modes should be investigated. From this point of view, an advanced methodology is not always required but rather typically depends on characteristics of the target structure.

Concrete sleepers typically display a characteristic in which a singularly damaged sleeper does not usually prompt a major impact on train-running safety or riding-comfort provided they can transmit trainloads and retain a gage [3]. Only in instances of multiple and consecutive concrete sleepers maintaining serious damage levels are the potential impacts on safety or comfort. This means that a state evaluation method based on continuous monitoring or advanced detection techniques is not necessary for concrete sleepers. If one can ascertain whether or not a measured concrete sleeper needs replacement, it could sufficiently contribute to the achievement of an effective maintenance protocol for them even if the utilized detection methodology is widely regarded as antiquated.

Several related references have previously pointed out the possibility of detecting concrete sleeper damage based on measured modal characteristics. Lam et al. [17, 18] and Kaewunruen and Remennikov [19, 20] developed ballast damage detection methods using the natural frequency and mode shape of in-situ concrete sleepers. In those studies, frequency-based damage detection for concrete sleepers was described as "an important future task". In addition, Kaewunruen and Remennikov [21] investigated the effects of rail pad stiffness on the modal parameters of sleepers, and Matsuoka et al. also investigated the modal properties of damaged sleepers [22]. Despite these contributions, the overall feasibility of the damaged-sleeper detection process remains uncertain because of the absence of the following three critical factors: impact of

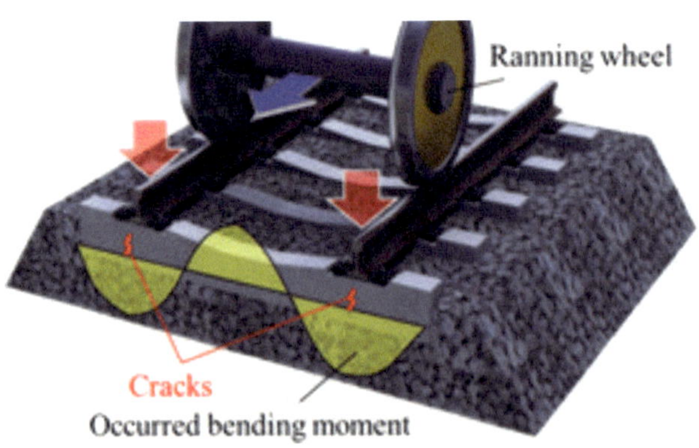

Figure 1.
Illustration of concrete sleeper installations and typical bending moment distribution during train passage.

concrete sleeper damage on modal characteristics, influences of specification variations of other track members, and a simple and efficient measurement process for practical use. An important contribution of this study is the provision of valuable data relating to impacts of typical damage modes, such as cracks and steel rod fractures, on the modal characteristics of general concrete sleepers in Japanese railway. Another contribution of this study is to propose a simple and efficient detection method assuming a practical application. As for vibration measurement methods of concrete sleepers, they have not been investigated other than via the use of accelerometers [19]. It is, however, difficult to apply such practical uses in which more than several tens of thousands of sleepers would require inspection per a rail line. This study therefore focuses on a detection method based on a hammering sound with a well-known impulse hammer technique to improve detection efficiency by omitting the installation of accelerometers.

2. Experimental methods

In this research, in order to verify the proposed damage detection system, several experiment series were set up, and necessary concrete sleepers were collected for each experiment. First series are new concrete sleepers. These are used to investigate the relationship between damage and mode characteristics through the stepwise bending test. These are also used to evaluate the influence of other track members (i.e., pad stiffness and ballast-supporting stiffness) on modal characteristics of sleepers on the actual environmental tests. Second series are concrete sleepers with actual damaged. Comparative studies of these concrete sleepers can provide an empirical validation of the feasibility of vibration-based damaged-sleeper detection. In addition, measurements using both accelerometers and sound-level meters were ultimately performed; comparison of such results can provide substantive evidence of applicability for the effectiveness of the sound-based detection method.

2.1 Test concrete sleepers

Figure 2 and **Table 1** provide design drawings and nominal specifications of test sleepers in this research endeavor. Specifically, this study focuses on 3PR and 3PO sleepers which are the most widely used types of sleepers on meter-gauged railway lines in Japan. 3PR/3PO are pre/posttension and mass/individual-production types, respectively. Posttensioning types utilize an unbonded system in which the steel-concrete bond is removed by an asphalt-based resin material.

Table 2 provides a list of test concrete sleepers. Six pretension and seven posttension type sleepers, which had been previously used in actual railway lines in Japan, were collected. Sleeper Nos. 1 and 9 shown as intact in **Table 2** were given artificial damage by stepwise bending tests, with vibration measurements. Sleeper Nos. 2–5 and 11–14 had different degrees of cracking or steel rod fractures generated through actual operational history. **Table 2** also presents these damage levels as "cracked sketches." Sleeper Nos. 6 and 13 are destroyed sleepers resulting from bending tests, in order to evaluate excessively damaged sleeper states. Other vibration measurements in a full-scale test line (described later) were performed on six intact sleepers (Nos. 17–22). In contrast, sleeper Nos. 7, 8, and 16 were used for verification of a simple and efficient detection method per the use of a sound-level meter.

2.2 Bending test method

Figure 3 presents the scheme of a bending test focusing on the cross-sections at the rail seats of Sleeper Nos. 1 and 9. This scheme complies with Japanese Industrial

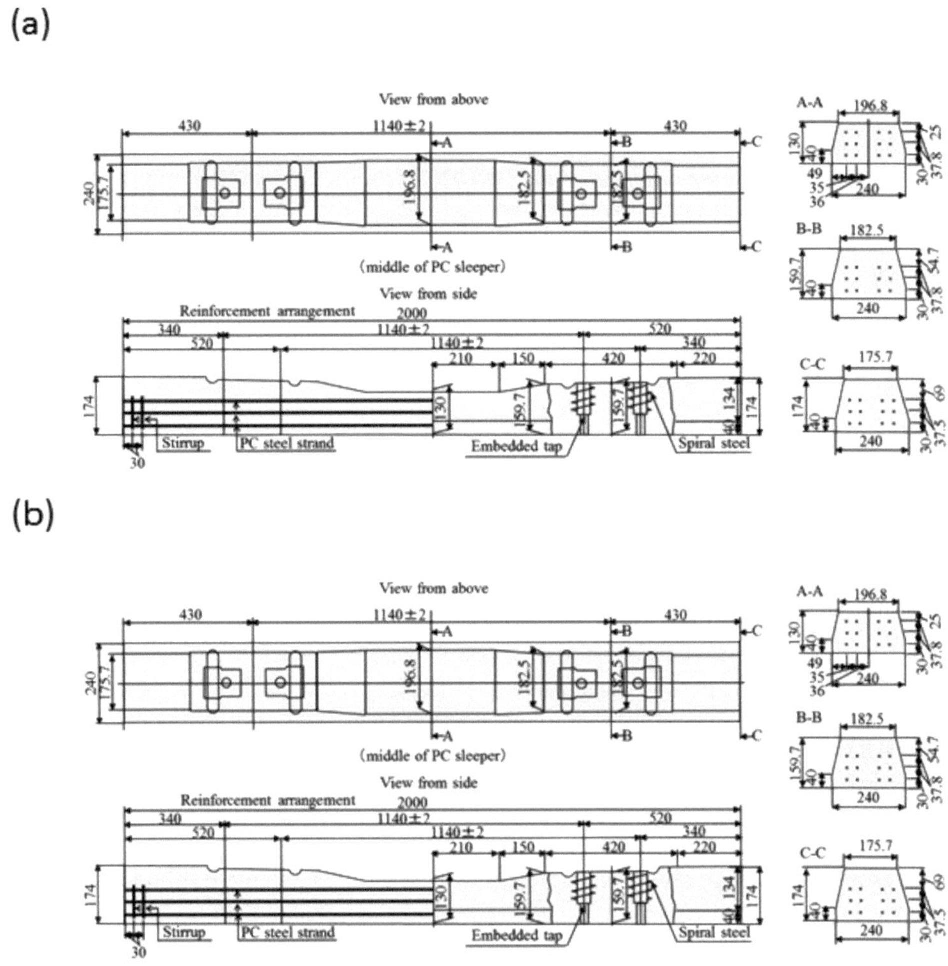

Figure 2.
Specifications of prestressed concrete sleeper on meter-gauged line prescribed in Japanese Industrial Standard: (a) the pretension type named 3PR and (b) the posttension type named 3PO.

3PR	Cross section	A-A	B-B
	Bending proof load	81 kN	50 kN
	Bending ultimate load	130 kN	86 kN
	Compressive strength	More than 49.1 N/mm²	
	Steel stranded wire	2.9 mm (3 stranded) × 12	
	Nominal elastic modulus of concrete	33.0 GPa	
	Concrete density	23.0 kN/m³	
3PO	Cross section	A-A	B-B
	Bending proof load	77 kN	45 kN
	Bending ultimate load	159 kN	92 kN
	Compressive strength	49.1N/mm² 以上	
	Steel lod	Nominal diameter 10 mm × 4	
	Nominal elastic modulus of concrete	33.0GPa	
	Concrete density	23.0kN/m³	

Table 1.
Nominal specifications of test concrete sleeper.

Standards E 1201 and 1202. In order to validate the impact of damage levels on modal characteristics, loading/unloading and vibration measurements were performed stepwise. Procedures for the bending test and the vibration measurement test are described as follows. First, a vibration measurement was performed in the intact state; second, vibration levels were measured after loading and unloading of the

Sleeper No.	Type	Test	Purpose	Damage level	Sleeper No.	Type	Test	Purpose	Damage level
1	3PR	I	Bending test	Intact	9, 10	3PO	I	Bending test	Intact
2	3PR	I	Actual damage	Crack at the bottom of RS (one side)	11	3PO	I	Actual damage	Crack at the bottom of RS (one side)
3	3PR	I	Actual damage	Cracks at the bottom of RS (both side)	12	3PO	I	Actual damage	Cracks at the bottom of RS (both side)
4	3PR	I	Actual damage	Cracks at the bottom of RS (both side)	13	3PO	I	Actual damage	Cracks at the bottom of both RS and the top of middle (shown as whole area cracks)
5	3PR	I	Actual damage	Cracks on whole area (Suspicion of ARS)	14	3PO	I	Actual damage	Cracks on whole area and a steel rod fractured
6	3PR	I	Ultimate state	After bending test of both rail and the middle position	15	3PO	I	Ultimate state	After bending test of both rail and the middle position
7	3PR	III IV	Sound measuring	Broken at the bottom of RS and cracks (one side)	16	3PO	III IV	Sound measuring	Cracks at the bottom of RS (both side)
8	3PR	III IV	Sound measuring	Intact	17–22	3PO	II	Other track member	Intact

* RS: Rail seat

Table 2.
List of collected concrete sleepers and associated damage levels.

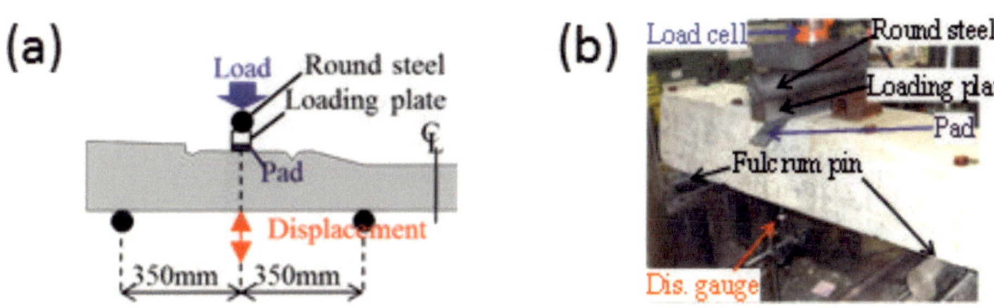

Figure 3.
Bending test scheme following the Japan Industrial Standards for concrete sleepers: (a) loading and supporting condition and (b) picture of installed concrete sleeper.

design-proof load; and third, after loading and unloading of the "cracking load," vibrational levels were again measured. The cracking load was determined by visual inspection which was conducted during the loading of this latter step. Subsequently, after loading and unloading 1.2 times the cracking load observed in the last step, the vibration level was again measured. Finally, after loading and unloading of the "ultimate load," the resulting vibration levels were once again determined. The "ultimate load" is defined as the maximum load. Bending tests with these procedures were conducted for both cross sections of rail seats for concrete Sleeper Nos. 1 and 9.

2.3 Vibration and sound pressure test method

Table 3 provides the specifications of measurement equipment associated with this study. In order to identify vibration characteristics up to 1 kHz, all acceleration and sound pressure responses were recorded on a Laptop PC with a 5 kHz sampling rate via a preamplifier and an A/D converter. Accelerometers and sound-level meters were employed (models PV85 and NL62, Rion Co. LTD.) that have frequency ranges up to 7 and 20 kHz, respectively. Sound pressures were measured by a sound-level meter with a flat input window. An impulse hammer was adopted per PCB PREZORTONICS INC. model 086C03 with a 2.25 mV/N sensitivity load cell to excite up to a level of 1 kHz. The weight of the unit was 136 g. **Figure 4(a)** shows the observed input forces of the impulse hammer and indicates an overall relative flat-frequency response. **Figure 4(b)** represents the distribution of all input forces, with the average of the forces being 0.079 N.

Figure 5 describes the four vibration test methods and shows TEST I using 11 accelerometers in order to validate the impact of artificial or actual damage on mode characteristics of sleepers. Each concrete sleeper was supported by a soft urethane mattress of 600 mm thickness to simulate a free-free boundary condition [23]. TEST I was conducted on Sleeper Nos. 1–7 and 9–13. **Figure 5(b)** presents TEST II using nine accelerometers for the purpose of estimating the influences of other track members. TEST II was performed in a test line on the premises of the Railway Technical Research Institute, in which sleepers with two rails and rail pads were laid on ballast similar to an actual railway environment. Concrete Sleeper Nos. 17–22, all which were intact, were evaluated in TEST II.

Figure 5(c) and **(d)** presents TEST III and IV using two and one accelerometers and a single sound-level meter to validate the feasibility of an effective detection method. These tests were conducted on Sleeper Nos. 7, 8 and 16. TEST III and TEST IV adopted the supporting method via the use of a soft urethane mattress and ballast, respectively. TEST III also investigated the impact of other vibration modes, excitation points, and sound-level meter distances to investigate the optimal measurement method and to clarify the potential limitations of usage in practical applications. In all tests, free-vibration responses excited by an impulse hammer were measured. Impact forces were applied at the midspan (and at the rail seat for TEST III) of sleepers, with responses being recorded three separate times for the respective sleepers.

Instruments	Model	Corporation	Specifications
Piezoelectric acceleration meter	PV85	Rion Co., LTD.	Frequency range: 1–7000 Hz Charge sensitivity: 6.42 pC/(m/s^2)
Sound-level meter	NL-62	Rion Co., LTD.	Frequency range: 10–20000 Hz Frequency characteristic: Flat
Pre-amplifier	NH-22	Rion Co., LTD.	Frequency range: 10–100000 Hz
Impulse hammer	086C03	PCB PREZORTONICS INC.	Frequency range: 0–8000 Hz Hammer mass: 136 g
Data acquisition system	Ni cDAQ-9172 Ni9233 LabVIEW	National Instruments Japan Corporation	Sampling frequency: 5000Hz

Table 3.
Specifications of measurement equipment.

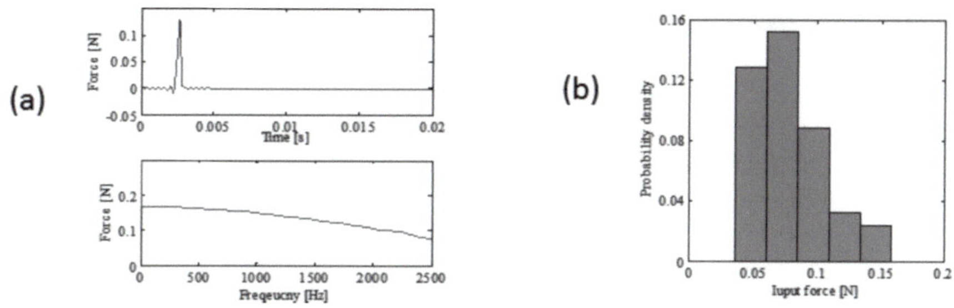

Figure 4.
Input force by impulse hammer: (a) time and frequency series and (b) distribution of input force.

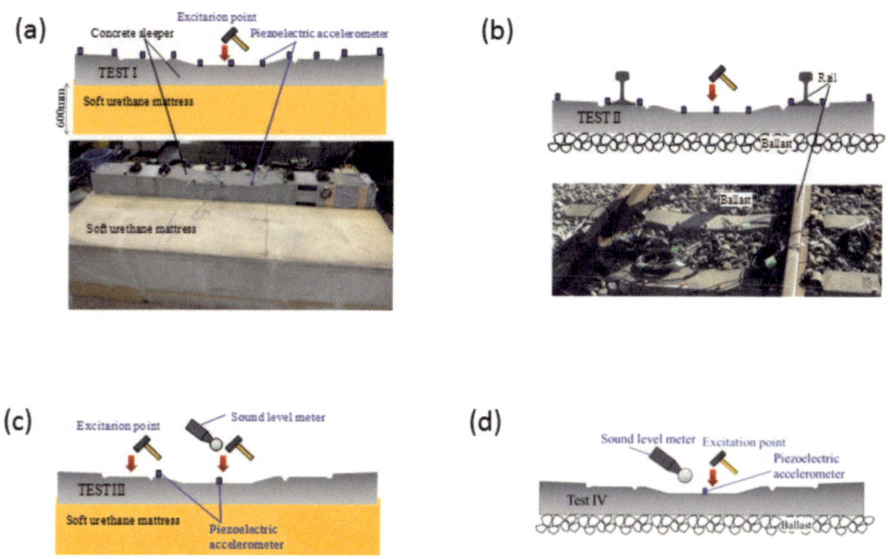

Figure 5.
Sensor arrangement and impact position of vibration and sound pressure test methods: (a) TEST I assuming free-free boundary condition; (b) TEST II assuming actual boundary condition; (c) TEST III assuming free-free boundary condition; and (d) TEST IV with ballast support.

2.4 Identification methods

Two well-known methods, ERA (Eigensystem Realization Algorithm) [23] and peak-picking were employed to identify the modal characteristics, natural frequency, modal damping ratio (ERA only), and modal shape (ERA only). Peak-picking was embedded into portable equipment for practical applications. Both identification methods, however, were neither capable of considering nonlinearity nor nonstationarity that may be caused by damage, such as breathing cracks. Several publications [24] have previously indicated that unsteady modal characteristics can express smaller structural damage scenarios. Concrete sleeper management, however, does not require the detection of such smaller-type damage events; being able to exclusively detect damaged sleepers that are in need of exchange is sufficient. Thus, this study adopted well-known time-invariant identification methods. As for ERA, more information about identification via the ERA method can be found in Refs. [25, 26]. Peak picking, which is the most simple and straightforward identification method for natural frequencies, was employed to develop an efficient detection system in the practical field. A 0.5-s acceleration just after impact excitation at the sleeper midspans was used for identification; thus, the resulting frequency increments were 2 Hz.

2.5 Numerical calculation method

In order to characterize the dynamics of concrete sleepers, many types of numerical models for concrete sleepers have been proposed. Grassie [5] proposed a simplified two-dimensional dynamic model in the free-free condition, with their analytical results having been verified by comparisons with 12 different types of sleepers. Dahlberg and Nielsen [27] developed a concrete sleeper model popularly denoted as the "Timoshenko Beam on an Elastic Foundation," for both free-free and in-situ conditions. Lam et al. [17, 18] also modeled the sleeper as a Timoshenko beam, the supporting ballast as discretized springs, and the rails as masses (with reference to previous studies [28, 29]). Furthermore, Kaewunruen and Remennikov [20] modeled the in-situ concrete sleeper as the sleeper, and the ballast and pads using the effective stiffness of the rails and pads.

In order to characterize affecting mechanisms of sleeper damage on modal characteristics, numerical calculations based on a finite-element model were performed in this study. **Figure 6** shows the subject numerical model for the concrete sleepers. Numerical analysis for the concrete sleepers was performed by LS-DYNA, version R8.0.0 [30]. A sleeper's concrete was modeled as hexahedral solid elements, and its steel wires and stirrups were modeled as beam elements. Supports and loading points for loading test analyses were modeled as rigid elements. A sleeper itself was modeled as a symmetrical model. Solid elements (concrete) and beam elements (steel wire and stirrups) share actual nodes to prevent slippage from occurring. **Table 4** shows the material properties of each element in the sleeper model. Young's modulus of concrete was set from the stress-strain curve. Uniaxial compression strength and uniaxial tensile strength of concrete was set from Young's modulus and Design Standards for Railway Structures and Commentary (Concrete Structures) [31]. For concrete, a material that can address cracking with tension softening and crushing was ultimately used [32]. Prestressing was reproduced by initial stress to the steel wires in the axial direction.

In order to reproduce the influence of bending damage upon modal characteristics, numerical simulations of loading and unloading under the same experimental bending conditions were first calculated, and then, the modal characteristics of damaged sleepers were investigated by eigenvalue analysis in each loading-unloading step.

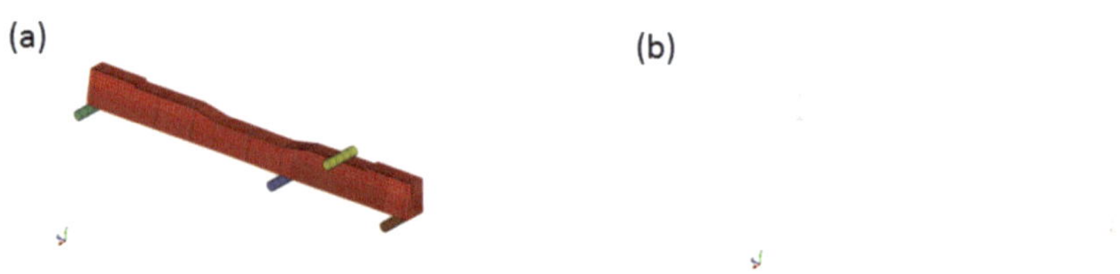

Figure 6.
Numerical model of a concrete sleeper: (a) finite-element of concrete and (b) reinforced steel.

Material	Element type	Young's modules (kN/mm^2)	Poisson's ratio	Density (kN/m^3)	Uniaxial compressive strength (kN/mm^2)	Uniaxial tensile strength (kN/mm^2)	Cross-sectional area (mm^2)	Initial tensile stresses (N/mm^2)
Concrete	solid	30	0.2	20	40	2.69	—	—
Steel wire (dia. 2.9 mm*3)	beam	200	0.3	77	—	—	19.82	1448
Strrup (dia. 4 mm)	beam	200	0.3	77	—	—	12.56	—

Table 4.
Element types and specifications of the numerical model.

3. Results

3.1 Bending test results

Figure 7 shows the load-displacement relationship at the rail seat, and **Table 5** shows applied loads of each state. In all cases, the cracking loads and ultimate loads were overdesign-proof and design-ultimate oriented. It can be confirmed that the bending of sleepers tends to soften gradually after the load exceeds cracking levels (i.e., around 110 kN). Both the cracking and ultimate loadings of Sleeper No. 9 in the posttension system are larger than those of Sleeper No. 1 in the pretension system. Although bending tests were performed at the right and left rail seats, explicit differences between the respective cases were not observed.

Figure 7, moreover, presents numerical calculation results. The results agree well with the experimental findings for pretension type Sleeper No. 1. A first crack occurred at the bottom of the rail seat as shown in **Figure 8**. The crack load is altogether depicted in **Table 5**, and the 112 kN cracking load agrees with the 111 kN

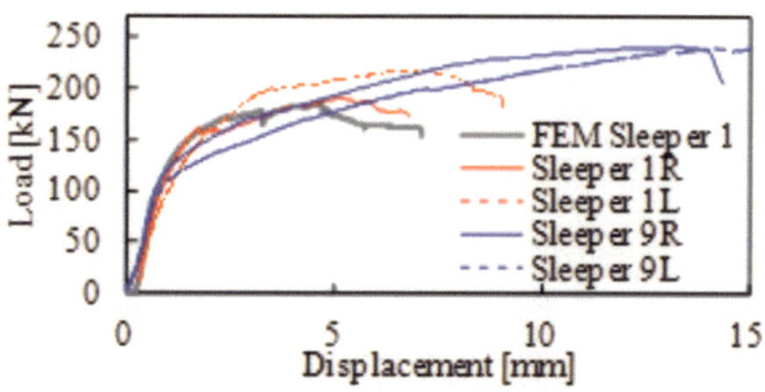

Figure 7.
Load-displacement envelope curve on bending test.

States	Applied loads [kN] 1R	1L	9R	9L	FEM
Undamaged	0	0	0	0	0
Design proof	81	81	77	77	81
Cracking	111	108	124	126	112
1.2 × cracking	133	130	149	151	134
Ultimate	190	215	240	245	184

(R: right side, L: left side)

Table 5.
Bending test results.

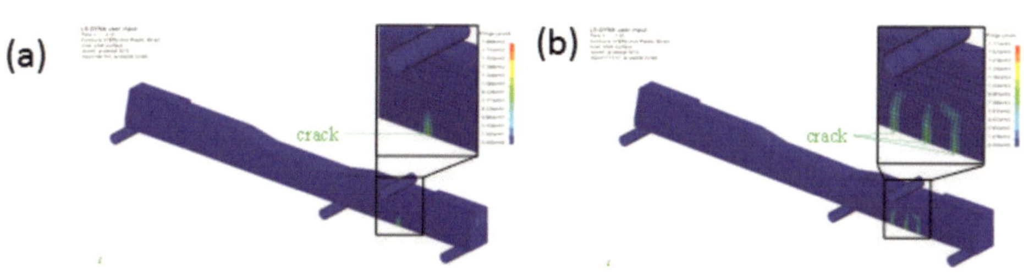

Figure 8.
Contours of cracks at (a) 112 kN and (b) 184 kN applied loads.

experimental result. For the ultimate state, three total cracks were found at the rail seat. This fracture mode is likewise consistent with the experimental one, as shown in **Figure 9**. Thus, it can be confirmed that the numerical model can accurately reproduce those realistically propagated in the damaged state.

Figure 10 shows the crack-width at the bottom of the rail seat calculated by the numerical model. **Figure 10(a)** indicates the general trend of prestressed concrete and that the maximum width of the first crack, crack 2, becomes wider with increasing load. In contrast, **Figure 10(b)** shows the residual crack-width after unloading. Residual cracks first appeared after an approximate 170 kN loading, which is close to the ultimate load. This means that most opened cracks due to such a loading can ultimately close back together after unloading, due to residual prestresses.

3.2 Modal identification results with bending tests

Figure 11 presents the time history and Fourier spectrum of measured accelerations at Sleeper No. 1 prior to the bending test. The vibrational measurements of Sleeper Nos. 1 and 9 (intact) were conducted based on TEST I, and the modal characteristics were identified by the ERA method.

Figure 12 shows the identified and numerically calculated natural frequencies and modal shapes from the first to third modes of Sleeper No. 1. These numerical results were calculated using the calibrated numerical model. The model displayed a value of 1.3 times that of the nominal elastic modulus for concrete based on the core

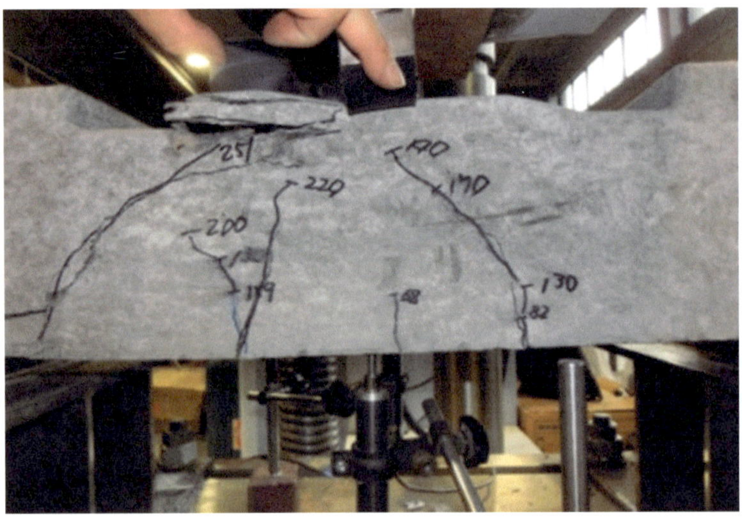

Figure 9.
Photograph of cracking at ultimate condition.

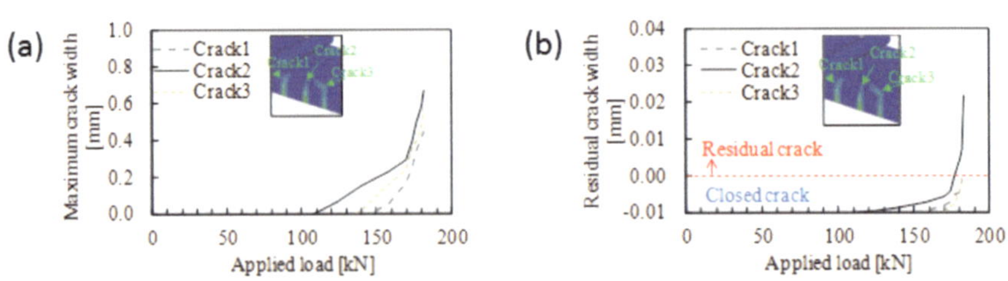

Figure 10.
Relationship between maximum applied load and crack width: (a) maximum crack width under load and (b) residual crack width after unloading.

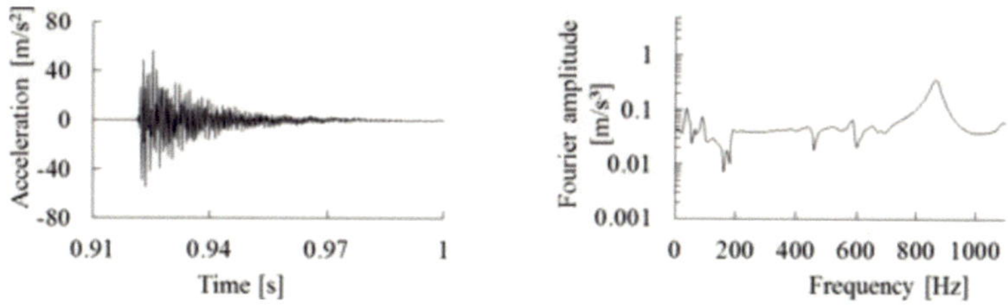

Figure 11.
Measured acceleration time history and spectrum on the midspan of Sleeper No. 1.

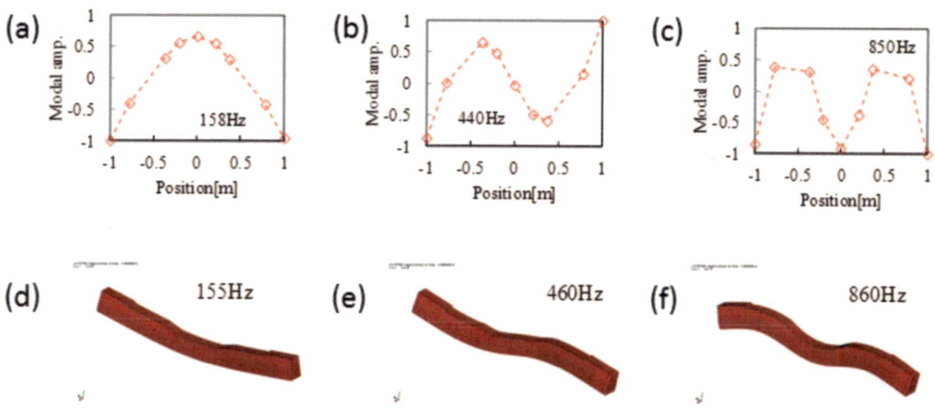

Figure 12.
Comparison of identified and calculated natural frequencies and mode shapes: (a)–(c) identified first-third bending mode; (d)–(f) calculated first-third bending mode of Sleeper No. 1.

test result of Sleeper No. 1. The numerical and identified natural frequencies and modal shapes showed good agreement within each mode.

3.3 Impact of applied loads on modal characteristics

Figures 13–15 show natural frequencies, modal damping ratios, and mode shapes of the first to third modes. These were identified by the ERA method using TEST I, which were conducted for each loading-unloading step. Numerical results, natural frequencies, and modal shapes, obtained by eigenvalue analysis, are shown in **Figures 13** and **15**.

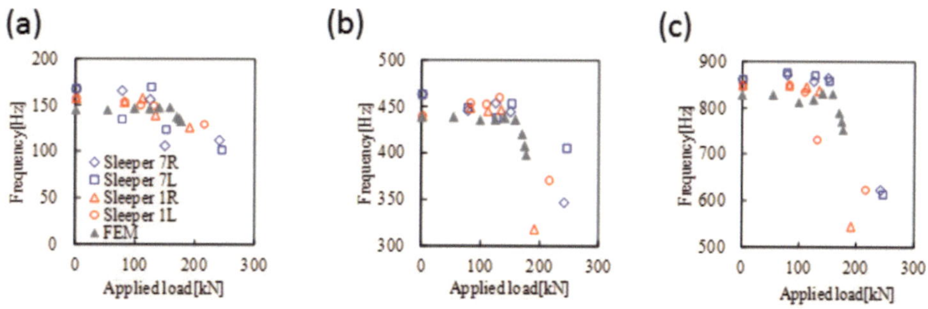

Figure 13.
Impact of applied load on natural frequency: (a) first bending mode; (b) second bending mode; and (c) third bending mode.

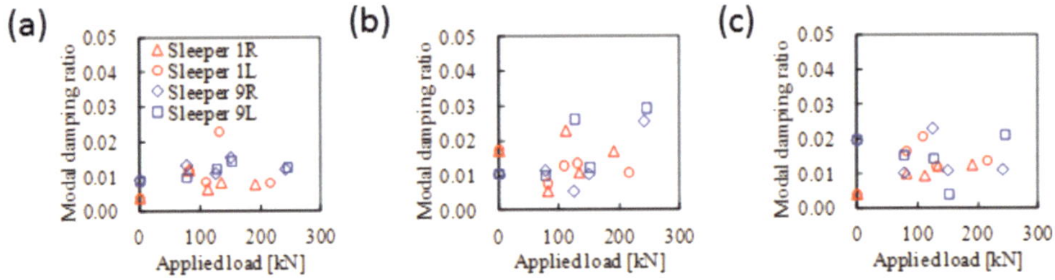

Figure 14.
Impact of applied load on modal damping ratio: (a) first bending mode; (b) second bending mode; and (c) third bending mode.

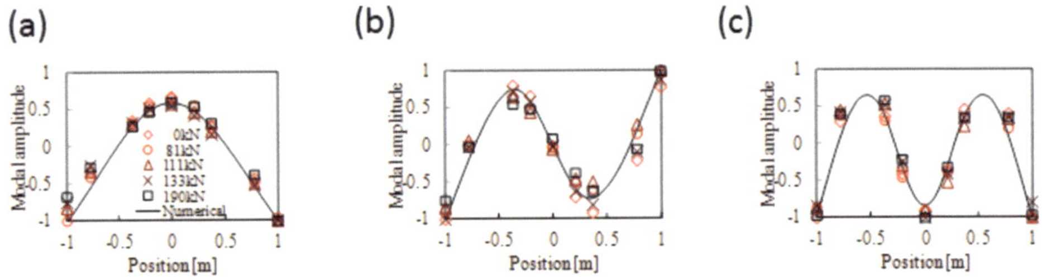

Figure 15.
The impact of applied load on right rail seat (−0.45 m position in each figure) and modal shape: (a) first bending mode; (b) second bending mode; and (c) third bending mode.

Figure 13 indicates that the natural frequency decreases rapidly when the applied load reaches a value of 1.2 times the cracking load (about 130 kN) or greater. This tendency is consistent with the numerical model when a concrete crack remains open even after unloading. Thus, it is perceived that a reduction in bending rigidity due to an open crack has a greater impact on a decrease in natural frequency than other bending-damage modes such as the plasticizing of steel rods or stranded wires, or reduction of introduced pretension. As shown in **Figure 13**, the natural frequency of the third mode in the ultimate state drops by approximately 150 Hz (17%) in all samples. It can therefore be asserted that the agreement of location between a crack generation and a higher modal amplitude might cause these large decreases in natural frequency within the third mode.

Figure 14 represents well-known large variations in mode-damping ratios. Some increasing tendencies of damping ratios due to increased applied loads can be seen in the first and second modes of Sleeper No. 9. Modal damping ratios are, however, difficult to apply as damage detection indicators due to a lack of certainty.

Figure 15 shows the modal shapes of Sleeper No. 1 from bending tests conducted at the left rail seat. It is typically difficult to find the impact of loading influences on modal shapes. If a modal shape is ultimately adapted as a detection indicator, another challenge would be at hand in which the use of numerous sensors would be required for carrying out multipoint measurements.

Figure 16 shows identified natural frequencies of concrete sleepers, in which damage has been generated during actual service. Vibration measurements were performed via TEST I on Sleeper Nos. 1–6 and 9–15 as shown in **Table 1**. Results of Sleeper Nos. 1 and 9 are representative of intact states prior to bending tests.

For the first mode frequency, some of the sleepers (e.g., Nos. 6 and 9) differ with the overall trend of decreasing frequency due to damage. The bending-tested Sleeper No. 6 and the intact Sleeper No. 9 have relatively high and low frequencies, respectively. Accordingly, it is not easy to detect damaged sleepers using first mode frequencies. With respect to the second mode, the frequencies of damaged sleepers

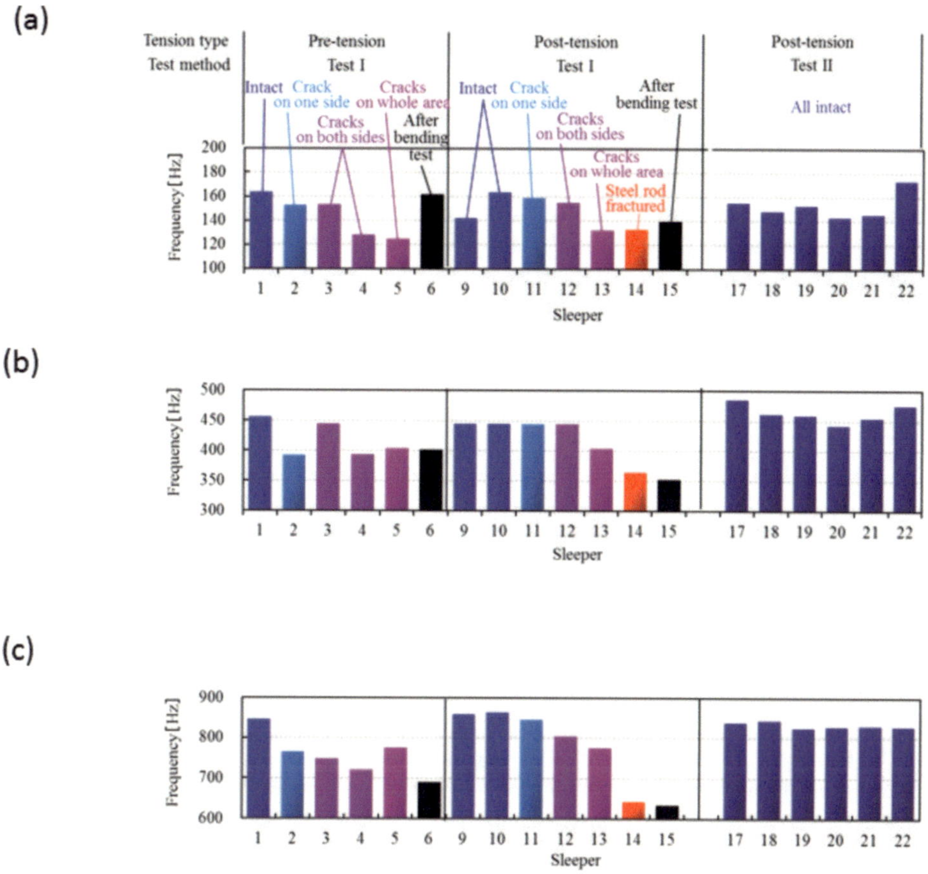

Figure 16.
Impact of actual damage and actual environment on natural frequency: (a) first bending mode; (b) second bending mode; and (c) third bending mode.

are low in comparison to those of intact ones. The natural frequencies of Sleeper No. 14 with a steel rod fracture and Sleeper Nos. 6 and 15 after postloading are significantly reduced. For the third mode, frequencies of damaged sleepers are clearly less compared with those of intact ones. In essence, the overall tendency seems to point to the notion that severity of damage typically corresponds to a proportional decrease in frequency. An inconsistency in this trend can be seen only in Sleeper No. 5 which has cracks in nearly all areas and yet exhibits a slightly higher frequency than that of less-damaged Sleeper No. 2. Sleeper Nos. 2 and 5, however, show sufficient declines in frequency to be able to clearly identify their relative levels of damage.

Thus, there were no hindrances to detect damaged sleepers when third bending frequencies were adopted as a detection index. Frequency-based damage detection is a well-known method based on simple concepts carried out through numerical analysis and laboratory tests; there are however, only limited examples of such application achievements within the civil structural domain. In response to this notion, this subject study strives to empirically validate the feasibility of a detection method using third bending mode frequency via experimental evaluation of actual damaged sleepers.

3.4 Impact of the actual environment on natural frequency

For operational railway tracks, concrete sleepers are laid on supporting ballast, along with two rails and pads. In particular, ballast-supporting stiffness has been historically measured in large variations [33]. Thus, a superior detection indicator should not only be sensitive to damage but also "insensitive" to the states of other track members such as ballast and pads. In order to validate the feasibility of frequency-based damaged-sleeper detection in the actual field, potential impacts from the external environment were investigated. Vibration measurement TEST II and modal identification were hence conducted on intact concrete Sleeper Nos. 17–22, which were on a test line within Railway Technical Research Institute premises.

Figure 16 presents the identified natural frequencies of Sleeper Nos. 17–22. A large variation could be confirmed in the first and second bending modes. It can thus be asserted that the variation of specifications in other track members causes this large variation because completely intact concrete sleepers themselves all display the same properties. On the other hand, the variations in the third bending mode are small. These empirical results are consistent with the trends pointed out in the existing literature [22], in which the variation of ballast-supporting stiffness mainly affects low-order modes, such as first and second bending. These results therefore imply that the third bending mode frequency is a suitable detection indicator, which consistently exhibits desirable characteristics for efficient damaged-sleeper detection, as described above.

4. Discussion about the simplification

4.1 Applicability of peak picking

The applicability of peak picking, one of the simplest methods for identifying third bending mode frequencies, was investigated. A Fourier spectrum was created by a 0.5-s acceleration at the midspan of sleepers just after an impact excitation. Peak frequencies corresponding to the third bending mode were picked up from the range of 500–1000 Hz frequency according to the aforementioned results discussed in this study.

Figure 17 presents identified third bending mode frequencies by ERA and peak picking. Both sets of results show excellent agreement; thus, peak picking can effectively identify the third bending mode frequency (i.e., suitable detection indicator) based on a simple measurement using only a single accelerometer.

Figure 18 presents measured FFT spectra of three tests, which translates to a certain reliability level within the peak picking method. Peaks of three measured spectra were in good agreement for each test sleeper. It should be noted that only damaged Sleeper Nos. 3, 4, 12, and 14 have some minor peaks and discrepancies at frequency ranges other than the peaks between test cases. This complex influence might be caused by nonlinearity or nonstationarity due to damage; thus, there is some degree

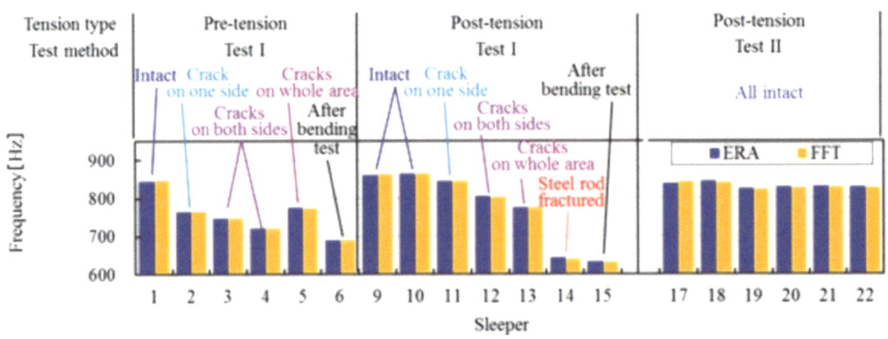

Figure 17.
Comparison of identification methods of third bending mode frequency based on TEST I and TEST II.

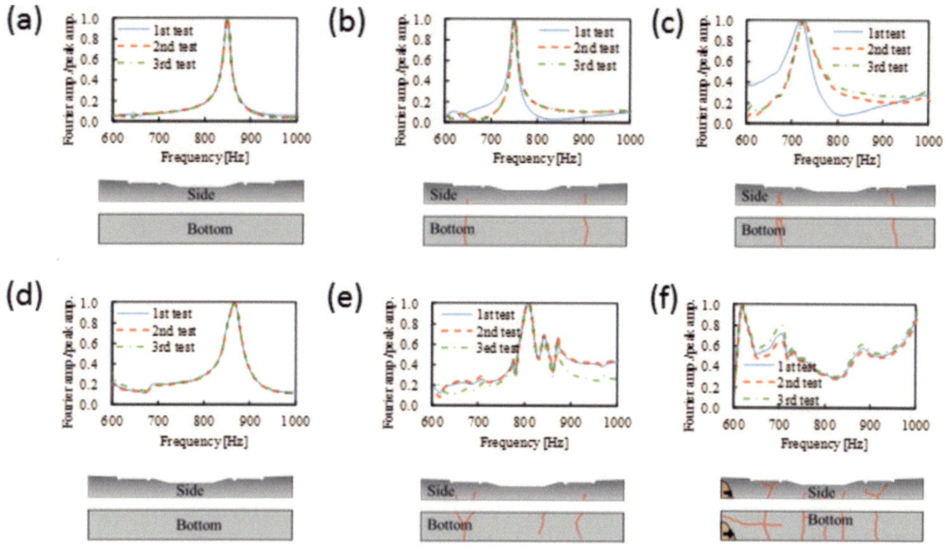

Figure 18.
Reliability of measured spectrum based on TEST I: (a) Sleeper No. 1; (b) Sleeper No. 3; (c) Sleeper No. 4; (d) Sleeper No. 9; (e) Sleeper No. 12; and (f) Sleeper No. 14.

of possibility that such damage can be detected by focusing on those influences. The peak frequency of concrete sleepers is, however, reduced because of damage and can be easily and reliably found by the peak picking method in the absence of information regarding nonlinearity or nonstationarity. Thus, it can be concluded that peak picking is sufficient for damaged concrete sleeper detection as maintained in this study.

4.2 Applicability of hammering sound

In the quest for additional simple measurement techniques, damage detection by sound-level meters that do not require the installation of accelerometers was experimentally investigated focusing on Sleeper Nos. 7, 8, and 16.

Structural vibrations can be propagated to peripheral regions as acoustic radiation via the air. Thus, sound pressure caused by impulse hummer test has possibility to be used for indirectly identification of concrete sleepers. Considering the modal characteristics and convenience of practical use, this study sets the sound-level meter position to above the midspan of the sleepers, which can match the antinode positions of the third bending modal shape. In addition, this feasibility study experimentally investigated the effect of each mode on measured sound pressure and then provided an optimized measurement method to ultimately obtain the third bending mode frequency of test sleepers to apply it for practical uses.

Figure 19 shows Fourier spectra of acceleration responses and sound pressure. The acceleration spectra at the midspan and near the rail seats of a sleeper, in addition to sound pressure, are depicted in the **Figure 19**. **Figure 19(a)** presents the results when an impulse hammer excitement occurs at the midspan, and **Figure 19(b)** corresponds to when the hammer is excited at the rail seat. **Figure 19** indicates good agreement between spectra of acceleration and sound pressure around the frequency peak corresponding to the third mode (750–800 Hz). This fact supports the applicability of sound pressure measurement as a robust tool for damaged-sleeper detection. Another peak of sound pressure(s) in the realm of 350–400 Hz corresponds to the accelerations not at the midspan, but exclusively at the rail seat. Thus, these peaks are caused by second bending vibrations of sleepers. The sound pressure can thus surveil not only the third mode but also the second mode. However, when comparing

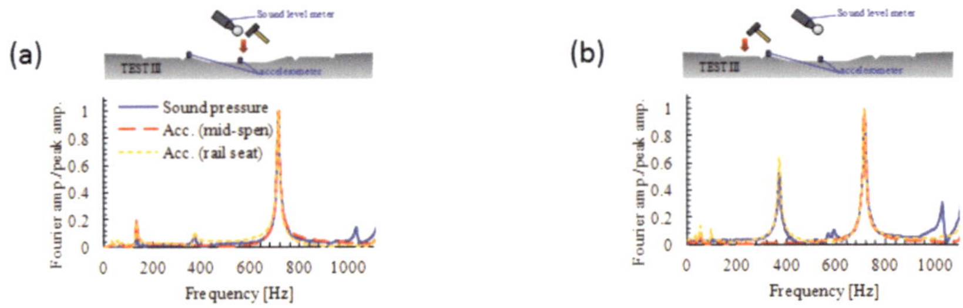

Figure 19.
Comparison between acceleration and sound pressure based on Fourier spectrum in TEST III: (a) Sleeper No. 7 with midspan excitation and (b) Sleeper No. 7 with rail seat excitation.

between midspan and rail seat excitation, the midspan excitation can ultimately reduce the second mode vibration level and hence make it easier to identify the third mode frequency by peak picking. It should be noted that the first mode (150–200 Hz) has little impact on the sound pressure because the vibration magnitude is significantly smaller than those of the second and third modes.

Figure 20(a) and **(b)** shows the impulse hammer test scheme to investigate the influence of sound observation positions. As shown in **Figure 20(a)**, the position of the sound-level meter was varied among 0.1, 0.3, and 0.5 m from the top surface of the sleeper. In addition, a convenient method of wearing the sound-level meter around the worker's neck (as shown in **Figure 20(c)**) was performed. The peak frequencies of the third mode for Sleeper No. 16 were extracted by peak picking. **Figure 20(d)** shows the extracted peak frequencies of measured acceleration and sound pressure corresponding to the third mode. It was confirmed that the peak frequency of sound pressure can estimate the third mode at the same value for all positions in this test and that these were consistent with the peak frequencies of

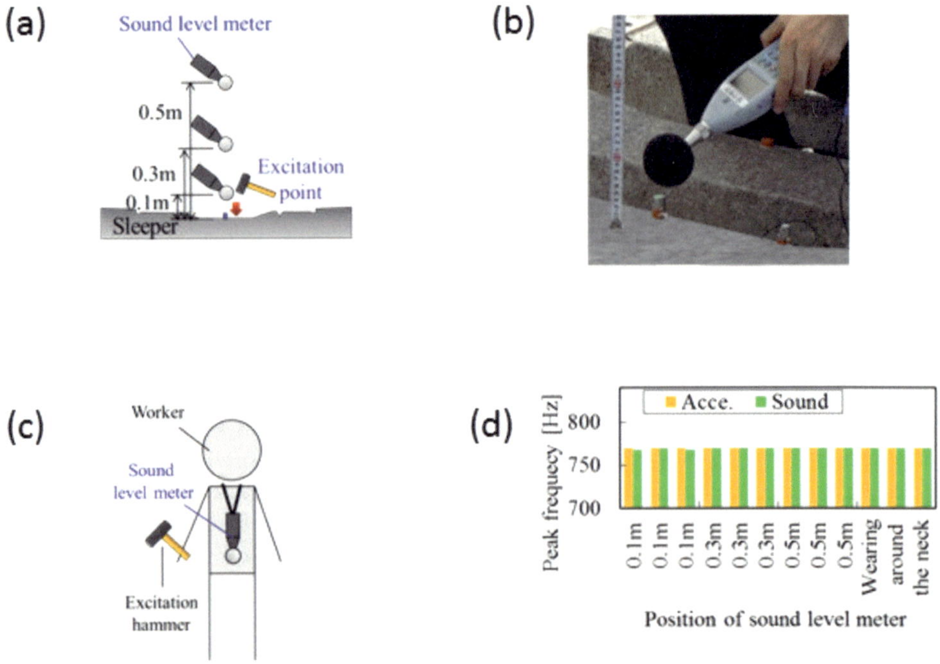

Figure 20.
Influence of distance of sound-level meter from excitation point: (a) test position; (b) example of sound-level meter condition; (c) wearing around the worker's neck and (d) comparison of peak frequencies.

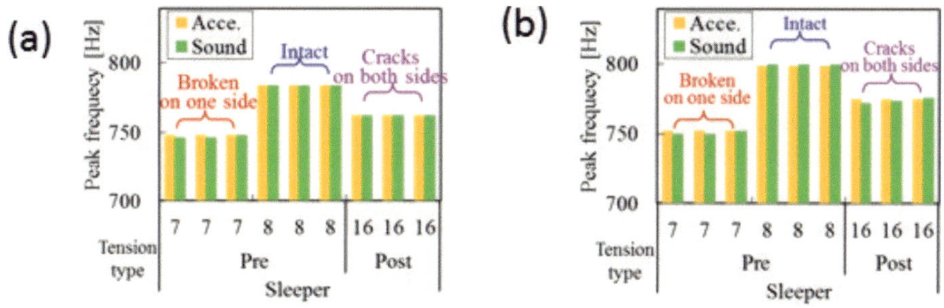

Figure 21.
Peak frequencies of acceleration and sound pressure based on (a) TEST III with support at both ends and (b) TEST IV with ballast support.

acceleration at the midspan. Thus, appropriate positioning of the sound-level meter is required to set it in the vicinity of the excitation point. **Figure 20(d)** illustrates that the same peak frequencies are obtained by a worker who performs an impulse excitation measurement by wearing a sound-level meter about the neck. This demonstrates an efficient damaged-sleeper detection protocol that can excite, measure, and provide a determination all via a single worker.

Figure 21(a) and **(b)** summarizes the peak frequencies of acceleration and sound pressure obtained based on TEST III and TEST IV, which were performed on concrete Sleeper Nos. 7, 8, and 16. The excitation points were at the sleeper midspans, and sound-level meters were worn around the worker's neck. Peak frequencies of acceleration and sound pressure were all found to be in good agreement. In addition, the peak frequency of damaged Sleeper No. 7 was clearly less than that of intact Sleeper No. 8 in both TEST III and IV. Therefore, it can be empirically verified that induced damages decrease frequencies within the third mode and that such frequencies can be accurately estimated via sound pressure measurements, even if the supporting method is changed to ballast (Test IV) from a soft urethane mattress.

5. Conclusions

In order to validate the feasibility of a frequency-based damage detection method, which is a well-known concept but has seen minimal practical application within the realm of railway concrete sleepers, this study experimentally investigated the impacts of artificial or actual damage on the modal characteristics of such sleeper. In addition, an efficient detection method based on sound pressure and its applicability for practical use was empirically validated. The associated resulting conclusions are summarized as follows:

1. Based on vibration measurements performed in parallel with bending tests, it was confirmed that natural frequencies start to be reduced when greater than 1.2 times the cracking load is applied.

2. Via numerical study, it was confirmed that reductions in natural frequency are caused by open cracks which remain open after unloading has occurred.

3. It was verified empirically that the natural frequency of the third mode (which is not only sensitive against damage but also less influenced by pad stiffness and ballast-supporting stiffness) is a suitable indicator for damaged-sleeper detection based on acceleration measurement tests of sleepers with actual damage and measurement tests on full-scale test lines.

4. Natural frequencies of third mode suitable for damage detection coincide with the peak frequencies of acceleration at the central sleeper segments, and, the peak frequency of acceleration matches the peak frequency of sound pressure measured by a noise meter. Thus, it became clear that it is possible to detect damaged sleepers via a noncontact measurement mode per the use of a sound-level meter.

5. As the results of investigation for the noncontact measurement method inherently assume a practical application, the protocol of wearing a sound-level meter around a worker's neck could suitably estimate the frequency of the third mode for concrete sleepers and altogether enable the efficient detection of damaged sleepers, including the likes of excitation, measurement, and determination by a singular worker.

When comparing the natural frequencies of the third mode obtained by the noncontact measurement techniques presented in this study, many deep damaged concrete sleepers can ultimately be detected from a large number of those measured. Furthermore, the proposed process can significantly contribute in the determination of sleeper-exchanging priorities according to the amount of measured frequency reduction in various cases.

Conflict of interest

There are no conflicts of interest to declare.

Author details

Kodai Matsuoka* and Tsutomu Watanabe
Railway Technical Research Institute Structural Mechanics in Railway Dynamics Division, Kokubunji-shi, Japan

*Address all correspondence to: matsuoka.kodai.13@rtri.or.jp

References

[1] FIP Commission on Prefabrication, FIP Commission on Prefabrication. Working Group on Concrete Railway Sleepers, Fédération Internationale de la Précontrainte. Working Group on Concrete Railway Sleepers. Concrete Railway Sleepers. Thomas Telford Services Ltd. One Great George Street, London, United Kingdom; 1987

[2] Remennikov A, Kaewunruen S. Experimental investigation on dynamic railway sleeper/ballast interaction. Experimental Mechanics. 2006;**46**(1):57-66. DOI: 10.1016/s0014-5793(01)03293-8

[3] Wakui H, Okuda H. A study on limit state design method for prestressed concrete sleepers. Doboku Gakkai Ronbunshu. 1997;**1997**(557):35-54. DOI: 10.2208/jscej.1997.557_35

[4] Yun WY, Ferreira L. Prediction of the demand of the railway sleepers: A simulation model for replacement strategies. International Journal of Production Economics. 2003;**81**:589-595. DOI: 10.1016/S0925-5273(02)00299-2

[5] Grassie SL. Dynamic modelling of concrete railway sleepers. Journal of Sound and Vibration. 1995;**187**(5): 799-813. DOI: 10.1006/jsvi.1995.0564

[6] Gustavson R. Structural Behaviour of Concrete Railway Sleepers. Göteborg, Sweden: Chalmers University of Technology; Sweden; 2002

[7] Kaewunruen S, Remennikov A. Post-Failure Mechanism and Residual Load-Carrying Capacity of Railway Prestressed Concrete Sleeper Under Hogging Moment. North Melbourne, Australia: The Institute of Materials Engineering Australasia; 2006

[8] Kaewunruen S, Remennikov AM. Dynamic crack propagations in prestressed concrete sleepers in railway track systems subjected to severe impact loads. Journal of Structural Engineering. 2009;**136**(6):749-754. DOI: 10.1061/(ASCE)ST.1943-541X.0000152

[9] Dukkipati RV, Dong R. Impact loads due to wheel flats and shells. Vehicle System Dynamics. 1999;**31**(1):1-22. DOI: 10.1076/vesd.31.1.1.2097

[10] Barke D, Chiu WK. Structural health monitoring in the railway industry: A review. Structural Health Monitoring. 2005;**4**(1):81-93. DOI: 10.1177/1475921705049764

[11] Salawu OS. Detection of structural damage through changes in frequency: A review. Engineering Structures. 1997;**19**(9):718-723. DOI: 10.1016/S0141-0296(96)00149-6

[12] Johnson EA, Lam HF, Katafygiotis LS, Beck JL. Phase I IASC-ASCE structural health monitoring benchmark problem using simulated data. Journal of Engineering Mechanics. 2004;**130**(1):3-15. DOI: 10.1061/(ASCE)0733-9399(2004)130:1(3)

[13] Sohn H, Farrar CR, Hemez FM, Shunk DD, Stinemates DW, Nadler BR, et al. A Review of Structural Health Monitoring Literature: 1996-2001. USA: Los Alamos National Laboratory; 2003. DOI: 10.1.1.729.3993

[14] Rice JA, Spencer BF Jr. Flexible Smart Sensor Framework for Autonomous Full-Scale Structural Health Monitoring. Newmark Structural Engineering Laboratory. Champaign, IL, USA: University of Illinois at Urbana-Champaign; 2009. DOI: 10.12989/sss.2010.6.5_6.423

[15] Siringoringo DM, Fujino Y. Observed dynamic performance of the Yokohama-Bay bridge from system identification using seismic records. Structural Control and Health

Monitoring. 2006;**13**(1):226-244. DOI: 10.1002/stc.135

[16] Dilena M, Morassi A. Dynamic testing of a damaged bridge. Mechanical Systems and Signal Processing. 2011;**25**(5):1485-1507. DOI: 10.1016/j.ymssp.2010.12.017

[17] Lam HF, Hu Q, Wong MT. The Bayesian methodology for the detection of railway ballast damage under a concrete sleeper. Engineering Structures. 2014;**81**:289-301. DOI: 10.1016/j.engstruct.2014.08.035

[18] Lam HF, Wong MT, Yang YB. A feasibility study on railway ballast damage detection utilizing measured vibration of in situ concrete sleeper. Engineering Structures. 2012;**45**:284-298. DOI: 10.1016/j.engstruct.2012.06.022

[19] Kaewunruen S, Remennikov A. Application of Vibration Measurements and Finite Element Model Updating for Structural Health Monitoring of Ballasted Railtrack Sleepers With Voids and Pockets. New York, United States: Mechanical Vibration: Measurement, Effect, and Control, Nova Science Publishers; 2009

[20] Kaewunruen S, Remennikov AM. Effect of improper ballast packing/tamping on dynamic behaviors of on-track railway concrete sleeper. International Journal of Structural Stability and Dynamics. 2007;**7**(01):167-177. DOI: 10.1142/S0219455407002174

[21] Kaewunruen S, Remennikov AM. Sensitivity analysis of free vibration characteristics of an in situ railway concrete sleeper to variations of rail pad parameters. Journal of Sound and Vibration. 2006;**298**(1-2):453-461. DOI: 10.1016/j.jsv.2006.05.034

[22] Matsuoka K, Watanabe T, Minoura S, Sogabe M, Omodaka A. Vibration modes of damaged PC sleepers and development of a simple damage detection method using sound level meter. Journal of Japan Society of Civil Engineers, Ser. E2 (Materials and Concrete Structures). 2018;**74**:158-175. DOI: 10.2208/jscejmcs.74.158

[23] Juang JN, Pappa RS. An eigensystem realization algorithm for modal parameter identification and model reduction. Journal of Guidance, Control, and Dynamics. 1985;**8**(5):620-627. DOI: 10.2514/3.20031

[24] Douka E, Hadjileontiadis LJ. Time–frequency analysis of the free vibration response of a beam with a breathing crack. NDT and E International. 2005;**38**(1):3-10. DOI: 10.1016/j.ndteint.2004.05.004

[25] Nagayama T, Abe M, Fujino Y, Ikeda K. Structural identification of a nonproportionally damped system and its application to a full-scale suspension bridge. Journal of Structural Engineering. 2005;**131**(10):1536-1545. DOI: 10.1061/(ASCE)0733-9445(2005)131:10(1536)

[26] Matsuoka K, Kaito K, Watanabe T, Sogabe M. Identification of dynamic properties of open-deck viaducts under passing train loads. In: Civil Engineering Topics. Vol. 4. New York, NY: Springer; 2011. pp. 155-162. DOI: 10.1007/978-1-4419-9316-8_13

[27] Dahlberg T, Nielsen J. Dynamic Behaviour of Free-Free and In-Situ Concrete Railway Sleepers. London, United Kingdom: Chalmers University of Technology, Solid Mechanics, Services Ltd; 1991

[28] Ford R. Modal analysis of a concrete railway sleeper. In: Research Note AVG/RN881122-1, School of Mechanical and Industrial Engineering. Australia: UNSW; 1988

[29] Vincent G. Modal analysis and numerical modeling of a concrete railway sleepers [Master of Engineering thesis]. Göteborg, Sweden: Department of Structural Engineering, Chalmers University of Technology; 2001

[30] Hallquist JO. LS-DYNA Keyword User's Manual 970. Livermore, CA, United States: Livermore Software Technology Corporation; 2007. pp. 299-800

[31] Design RS. Design Standards for Railway Structures and Commentary (Concrete Structures). Tokyo: Railway Technical Research Institute; 2007

[32] Grassl P, Xenos D, Nyström U, Rempling R, Gylltoft K. CDPM2: A damage-plasticity approach to modelling the failure of concrete. International Journal of Solids and Structures. 2013;**50**(24):3805-3816. DOI: 10.1016/j.ijsolstr.2013.07.008

[33] Esveld C. Modern Railway Track. 2nd ed. Delft, Netherlands: Delft University of Technology; 2001

Chapter 7

Nanotechnology and Development of Strain Sensor for Damage Detection

Yumna Qureshi, Mostapha Tarfaoui, Khalil K. Lafdi and Khalid Lafdi

Abstract

Composite materials, having better properties than traditional materials, are susceptible to potential damage during operating conditions, and this issue is usually not found until it is too late. Thus, it is important to identify when cracks occur within a structure, to avoid catastrophic failure. The objective of this chapter is to fabricate a new generation of strain sensors in the form of a wire/thread that can be incorporated into a material to detect damage before they become fatal. This microscale strain sensor consists of flexible, untwisted nylon yarn coated with a thin layer of silver using electroless plating process. The electromechanical response of this sensor wire was tested experimentally using tensile loading and then verified numerically with good agreement in results. This flexible strain sensor was then incorporated into a composite specimen to demonstrate the detection of damage initiation before the deformation of structure becomes fatal. The specimens were tested mechanically in a standard tensometer machine, while the electrical response was recorded. The results were very encouraging, and the signal from the sensor was correlated perfectly with the mechanical behavior of the specimen. This showed that these flexible strain sensors can be used for in situ structural health monitoring (SHM) and real-time damage detection applications.

Keywords: composites, structural health monitoring, flexible yarn, strain sensor, conductive film Ag-coating, electromechanical behavior

1. Introduction

Composite materials, despite having better physical properties, are not exempt from limitations and drawbacks [1–4]. The mechanisms of damage initiation and propagation leading to ultimate fracture in these materials are very complex but very well established [5–11]. Structural health monitoring (SHM) is a well-known technique to examine the mechanical behavior of the structure during operation and to avoid its sudden failure [12]. In situ real-time SHM has been used frequently for detecting different types of damages in materials such as corrosion, deformation, debonding/delamination, fiber cracking, thermal degradation, intralaminar cracking, etc. to ensure safe and durable service life of the structures [13–18]. So, vast research had been carried out during the past years to develop SHM sensors, and this development took place gradually over time from strain gages, fiber optic sensors, and piezoelectric sensors to microelectromechanical systems (MEMSs) [19–21]. But they all have some limitations such as strain gauges behave as defects or inclusions, fiber optic sensors require lot of instrumentation and data analysis, brittle material is used in manufacturing piezoelectric sensors, and MEMSs are manufactured at microscale, which makes the manufacturing process difficult [22–25].

Another class of sensors, known as textile sensors or flexible sensors, is the new focus of study for many researchers. These are conductive strain sensor wires/threads in which electrical resistance varies reversibly to applied stress [26–29]. They can be developed at nanoscale or microscale and can overcome the limitation of brittle behavior of conventional piezo-resistive strain gauges. So, it is essential to understand the mechanical behavior of these flexible sensors for better structural integrity and longer service life. Moreover, it is also important to understand the concept of computational modeling of these flexible yarns to model and analyze their behavior numerically. However, very less research has been conducted to use the concept of coated yarn as a flexible piezo-resistive strain sensor for structural health monitoring without jeopardizing the mechanical behavior of core material especially numerically. Different researchers had worked on numerical models and had used finite element analysis (FEA) to predict the mechanical behavior of yarn [30–32]. With the advancement of computer-aided design (CAD) and computer-aided engineering (CAE), it is possible to investigate the mechanical behavior of yarn using finite element modeling (FEM) [33]. Many CAD models of filaments, yarns, and fabrics have been developed by researchers with most of them related to geometrical modeling of yarns based on single line yarn path also known as pitch length [34–37]. Some researchers have attempted to overcome difficulties like small- and large-scale deformation, complex material properties, and 3D modeling [38]. Several analytical models had been established for the estimation of mechanical tensile performance of yarns. The tensile behavior of yarn, using force method, was first studied 90 years ago, which was then extended to examine the mechanical behavior of continuous filament yarns [39, 40]. Other than force method, energy method was used to study the continuous filament and to predict the whole stress-strain behavior in Tenasco yarn which was first proposed by Treloar and Riding [41]. Then, Riding and Wilson [42] extended this study and predicted the stress-strain relations for materials such as low-tenacity Terylene, Super Tenasco, and Nylon 6-6. Moreover, energy method was also used to study the tensile and torsional behavior of bulky wool single yarn [43]. Cartraud and Messager [44] studied the model of 1 + 6 (six cylindrical filaments were wrapped around a straight filament at core) stranded fibrous structure under tensile loading. Vassiliadis et al. [38] suggested a computational method to study the mechanical behavior of multifilament twisted yarn from 2 to 1200 filaments based on FEM. However, up to this date and to the best knowledge of the author, very limited or no research has been conducted to experimentally and/or numerically analyze a coated yarn and to study the electromechanical response of coated yarn-based wire models.

In this chapter, the overall objective is to fabricate a conductive wire that functions like a piezo-resistive strain gage while not jeopardizing the structural integrity of the composite and acting as a real-time sensor during the operating condition. This was achieved by using untwisted nylon yarn and depositing silver metal coating on its surface. Initially, experimental tests were conducted to quantify the electromechanical behavior of this detector and analyze its performance. Then, a numerical model was developed to validate this sensor design and confirm the reproducibility of results. Due to their extremely small size and large-scale integration degree, the sensors had the remarkable characteristics of light weight, flexibility in design, low power consumption and noise level, short response time, high reliability, and low cost. Once the experimental results were validated numerically, these microscale flexible sensor wires were incorporated in the composite specimen to demonstrate its SHM application. The specimens were then tested in standard tensometer machine mechanically while the electrical response was recorded, which correlated perfectly with the mechanical behavior of composite specimen. This showed that these flexible yarn wires can be used as piezo-resistive strain sensors for SHM applications of composite structures.

2. Sensor fabrication and characterization

Sensors were developed using untwisted nylon yarn and by depositing a silver (Ag) metal film on the surface of its filaments because even though nylon yarn behaved well mechanically, it was poor in electrical conductance. Thin film coating was applied to overcome the conductance issue without jeopardizing structural integrity of each material. Electroless plating process was used for this purpose, and the sensor specimens were characterized by the following dimensions: length 50 mm, diameter of yarn 225 μm, and coating thickness 1–2 μm. This thickness of the coating film is the best compromise between uniform thickness throughout the yarn and good conductive flexible coating. These dimensions were confirmed when scanning electron microscopy (SEM) of sample wires was performed for the characterization of samples after deposition of silver coating.

3. Experimental test setup for sensor wires

Uniaxial tensile testing machine and oscilloscope were used simultaneously to examine the sensing behavior of this flexible strain sensor wire. Three experimental

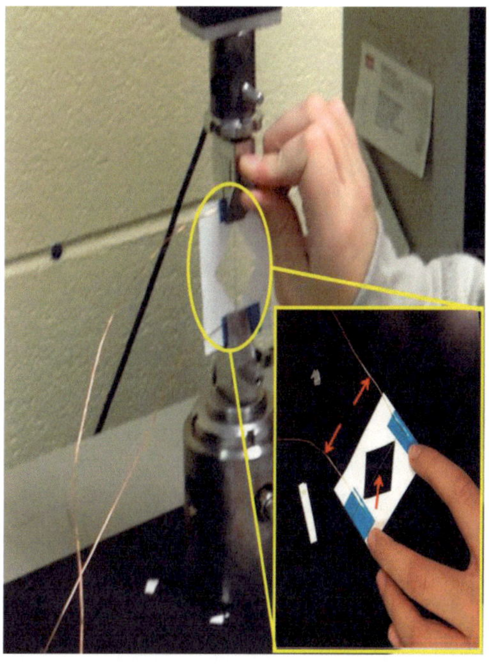

Figure 1.
Flexible sensor wire was attached with electrodes and paper support, placed in the tensile machine, and electrodes were attached to the data acquisition system.

tests were performed to confirm the reproducibility of the results. Paper frame was used specifically to provide support to such a small specimen in the tensile machine; however, it was cut from the sides before starting the test so it could not affect the behavior of sensor during the test. In addition, electrodes were attached at both ends of the specimen to provide better connection. Then, the specimen was placed in the tensile machine and test was performed at low strain rate, that is 5 mm/min, **Figure 1**. As a result, the stress-strain behavior with resistance profile was obtained.

4. Numerical modeling of sensor wire

Coupled field analysis in commercial Abaqus/standard software was used to model the electromechanical behavior and verify experimental results. The sample geometry of conductive thin film-coated monofilament was developed at the microscale to avoid aspect ratio problems during meshing and reduce the computational time, **Figure 2a** and **b**. Before performing the finite element analysis, it was

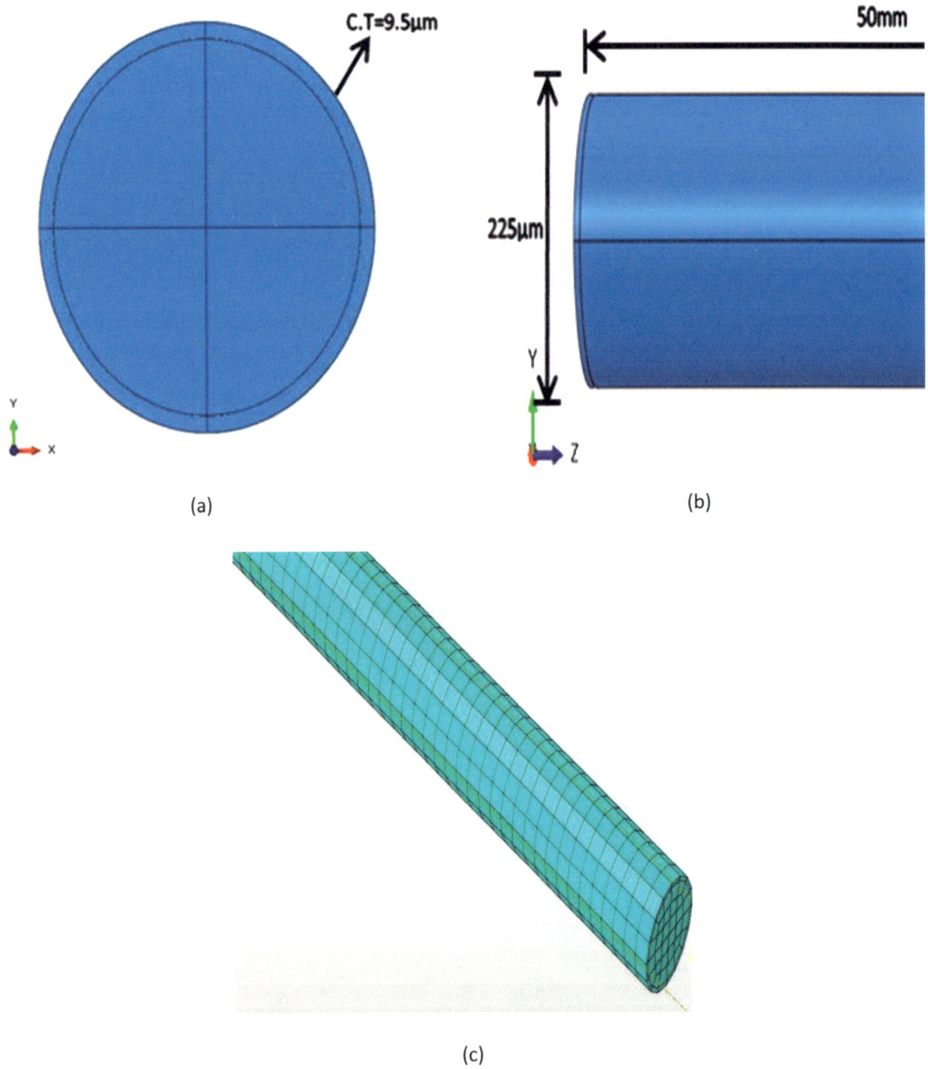

Figure 2.
Schematic representation of the coated yarn: (a) front view, (b) side view, and (c) mesh of the yarn.

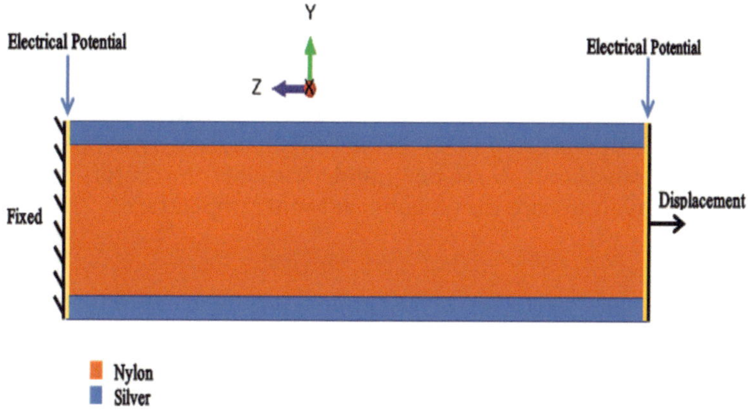

Figure 3.
Electromechanical boundary conditions applied to the 3D model.

important to ensure the convergence of the mesh. So, mesh convergence study was performed on the model to eliminate the dependency upon the mesh for which 5 mesh sizes, 1, 0.5, 0.1, 0.08, and 0.05 were considered, and, based on the study, mesh size 0.05 was used for discrete model because it was the best compromise between less computational work and accuracy of results, **Figure 2c**. The applied boundary conditions consisted of both mechanical and electrical loads, **Figure 3**.

5. Fabrication of composite specimens incorporated with sensor

Standard specimens of composites for tensile test were prepared and a full integration of sensors into the composite structure was achieved, **Figure 4**. Silicon

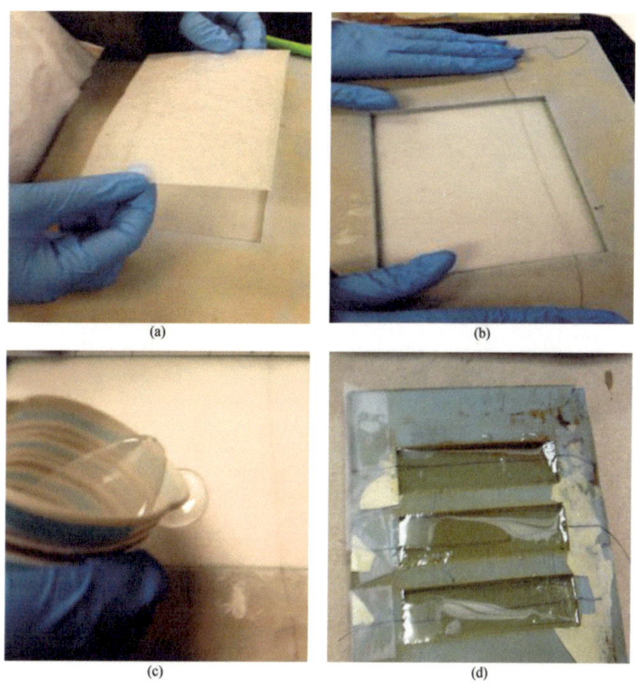

Figure 4.
Fabrication of composite specimen with sensor wire.

molds were used for the preparation of samples and glass fiber fabric was used as reinforcement. The glass fabric was cut into segments and placed into a mold and a sensor (nylon yarn coated with silver) was inserted between the plies of fiberglass. Then, resin mixed with hardener with a ratio of 1:4 was poured into the mold. Once the molds were filled, one could no longer see the fiberglass layers, the samples were completely transparent. Now, one could view all of the sensors easily. After that, samples were left to cure for 48 hours at room temperature.

6. Experimental test of composite with sensor wire

A cured composite sample with integrated sensor was tested using an Instron test machine and the oscilloscope was connected to sensor lids. Uniaxial tensile testing machine and oscilloscope were used simultaneously to correlate the mechanical behavior of the composite specimen with the electrical response of the strain sensor wire. Three experimental tests were performed to confirm the reproducibility of the results. In addition, electrodes were attached at both ends of the specimen to provide better connection. Then, the specimen was placed in the tensile machine and test was performed, **Figure 5**. As a result, the stress-strain behavior of composite specimen with resistance profile of the sensor was obtained.

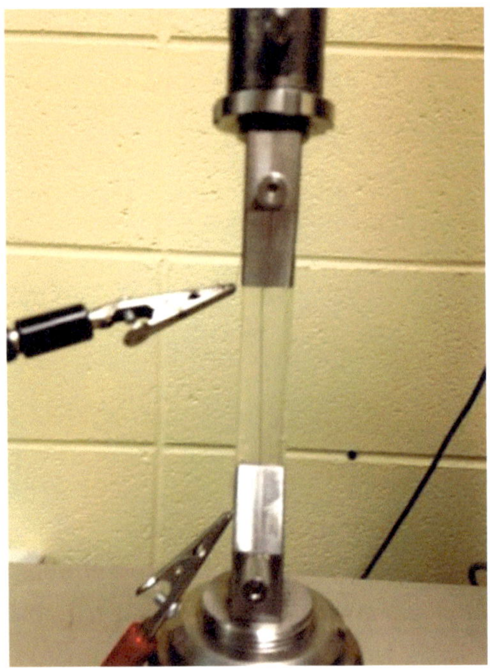

Figure 5.
Composite sample mounted on test frame for mechanical test. Sensor lids are connected to the oscilloscope.

7. Results and discussion

7.1 SEM characterization of flexible strain sensor wires

All the filaments were uniformly coated; however, after large magnification, it appeared that some filaments exhibited cracks or gaps. Another phenomenon observed during the SEM characterization was that the silver coating was made of

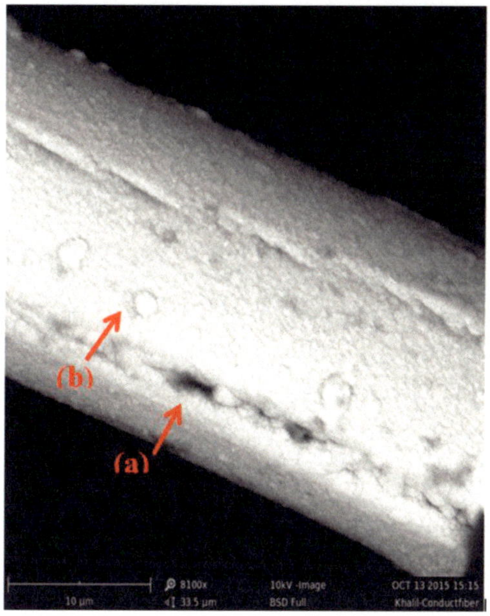

Figure 6.
SEM characterization of the Ag thin film-coated untwisted yarn: (a) shows the uniform coating with some minute discontinuities and defects in the coating and (b) confirms that the Ag thin film coating on the nylon surface consisted of deposition of nanoparticles of silver.

nanoparticle grains, or aggregates less than one micro-meter in size, **Figure 6**. SEM characterization of the coated untwisted yarn showed uniform coating with some minute discontinuities and defects and also confirmed that the Ag thin film coating consisted of deposition of nanoparticles of silver. Regardless of the imperfections in the Ag film coating in some filaments, it did not seem to affect the electrical conductivity.

7.2. Experimental testing

7.2.1 Electromechanical response

Three tensile tests of piezo-resistive sensor wire were performed. The Young modulus and yield strength of the tested samples were about 1348.5 and 20.13 MPa, respectively. The stress-strain behavior of untwisted coated yarn is shown in **Figure 7a**. Stress and electrical response of the untwisted yarn are plotted simultaneously in **Figure 7b**. The resistance was changed at the same time as the failure started to initiate, and, as the test progressed, the resistance increased gradually when the number of fractured filaments was increased. Ultimately, when the untwisted yarn was fractured completely, the resistance went to maximum value. The results were very encouraging, and piezo-resistive flexible sensor was responding very well to any change in load.

7.2.2 Damage modes

After tensile test, fractured samples were studied using SEM technique, and it appeared that there were two distinct morphologies. Some filaments exhibited a clean ductile failure in which both the coating and the fiber showed a clean cut; however, other filaments showed a pull-out of the coating or flaking off, **Figure 8**.

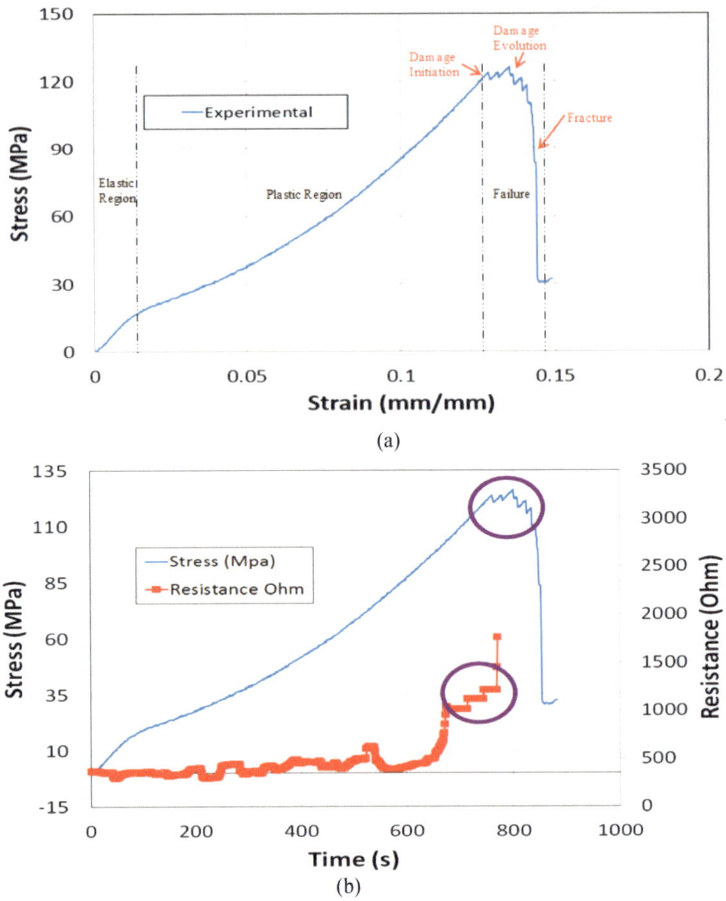

Figure 7.
Electromechanical behavior of untwisted nylon yarn: (a) mechanical behavior of untwisted nylon yarn and (b) electromechanical response.

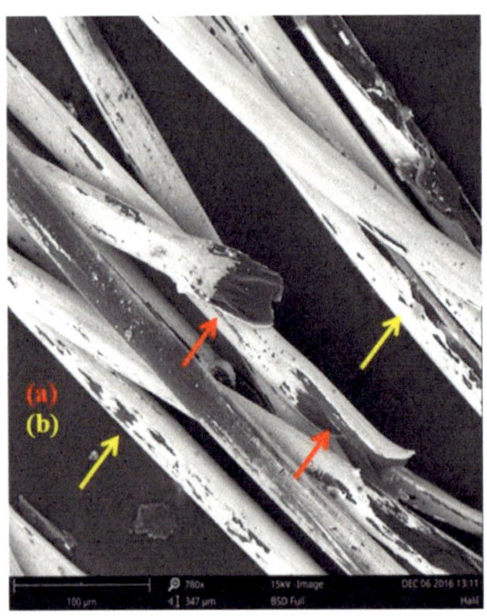

Figure 8.
SEM characterization of fractured samples confirms that the coating thickness was approximately 1–2 μm: (a) shows a clear ductile failure of both core and coating material and (b) depicts the second phenomenon, that is, flaking off of coating.

This established that some parts of the coating might not have developed strong adherence with the nylon fiber during the fabrication process. However, it did not affect the overall response of the sensor because of two reasons: first, the pull-out of coating happened during damage initiation and damage propagation just before the failure and did not affect the performance of the sensor before that; the second reason is that there were approximately 100 fibers in the single yarn and fiber pull-out happened in less than 5–10% of fibers, which is almost negligible. Nevertheless, the surface adhesion can be improved during the fabrication process by further improving the surface roughness by etching process before application of coating because rough surfaces have better adhesion properties.

7.3 FE analysis

The untwisted coated yarn was modeled as a ductile material using the built-in elastic, plastic, and ductile damage criteria of Abaqus because both silver and nylon are ductile in nature. Electrical conductance of both materials nylon and Ag thin film was defined in Abaqus to model the electrical response during the mechanical analysis, **Table 1**. For the numerical analysis, experimental tensile behavior of pure silver thin film was applied [45] in addition to the mechanical response of untwisted nylon yarn, **Table 1**.

Furthermore, the rate-dependent power law was defined using the experimental curves in the plasticity model because it plays a vital role in damage initiation and neck formation during ductile failure. Therefore, strain hardening stress coefficient K and strain hardening index n were calculated using Eqs. (1) and (2).

$$n = \frac{\log \sigma_2 - \log \sigma_1}{\log \varepsilon_2 - \log \varepsilon_1} \tag{1}$$

$$\log K - \log \sigma_1 = n(\log x - \log \varepsilon_1) \tag{2}$$

where $\sigma_{1,2}$ are stress points in the plastic region, $\varepsilon_{1,2}$ are the corresponding strain points in the plastic region, K is the strain hardening stress, and n is the strain hardening exponent.

Ductile damage criteria built in Abaqus was used to define model failure. Damage initiation depended on fracture strain, strain rate, and stress triaxiality whereas damage evolution required displacement at failure, **Table 1**. The evolution of the damage defined the material's behavior by illustrating the degradation of material stiffness after damage initiation. Scalar damage approach was used for formulating the rate of damage as given in (3). D is the overall damage variable showing the combined effect of all active damage mechanisms, and when it reached 1, fracture occurred.

$$\sigma = (1 - D)\,\acute{\sigma} \tag{3}$$

Material	Electrical conductance, S/mm	Young's modulus, MPa	Poisson ratio	Yield strength, MPa	Fracture strain	Strain rate, mm/min
Nylon	1×10^{-15}	1348.5	0.39	20.13	0.12	5
Silver	63×10^{3}	47,230	0.37	431.1	0.08	60×10^{-5}

Table 1.
Experimental elastic, plastic, and failure data of nylon and pure Ag-thin film.

where σ is the stress due to damage response, D is the damage variable, and $\acute{\sigma}$ is the stress due to undamaged response.

7.4 Verification of sensor response

The nylon monofilament coated with silver thin film was subjected to tensile elongation until failure. Results showed that it was viable to use one filament to validate the piezo-resistive behavior of untwisted coated yarn. The true stress-strain behavior showed a good correlation with the experimental results in the elastic-plastic region, **Figure 9**. It can be seen in the results that it was fine to use coated monofilament model to verify the result because the plane of stress is same. However, there is a slight difference in the failure initiation and breakage, which is understandable because: in experimental results, the failure shows gradual breakage of all the monofilaments, whereas in the numerical model, the set of monofilaments is modeled by a single thread. Electrical response was recorded as electrical current density (ECD) in Abaqus which was then converted to resistance response using Eqs. (4)-(6) to validate the experimental piezo-resistive behavior of sensor wire. Electromechanical behavior of the monofilament is shown in **Figure 10**.

$$J = \alpha E \ \ with \ \ \alpha = \frac{1}{\rho} \tag{4}$$

$$J = \frac{E}{\rho} \Rightarrow J \propto \frac{1}{\rho} \tag{5}$$

$$R = \frac{\rho L}{A} \tag{6}$$

where J is the current density, E is the electric field, α is the electrical conductivity, ρ is the resistivity, L is the length, A is the cross-sectional area, and R is the resistance.

It was observed that till the plastic region, the electrical resistivity of the yarn changed, but this change in resistance was very small as compared to change in resistance on damage when there was complete breakage in current flow. No gradual increase in the resistance was seen like in experimental results because of the monofilament model. The 3D discrete model of coated monofilament before and after failure is shown in **Figure 11**.

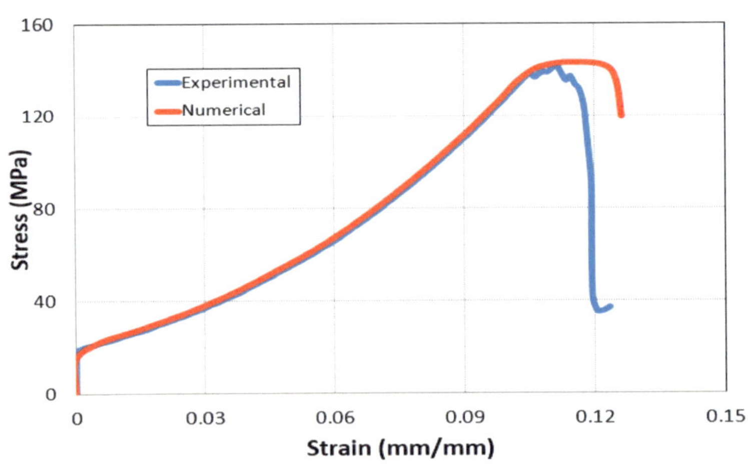

Figure 9.
Numerical verification of experimental mechanical behavior of Ag-coated untwisted nylon yarn.

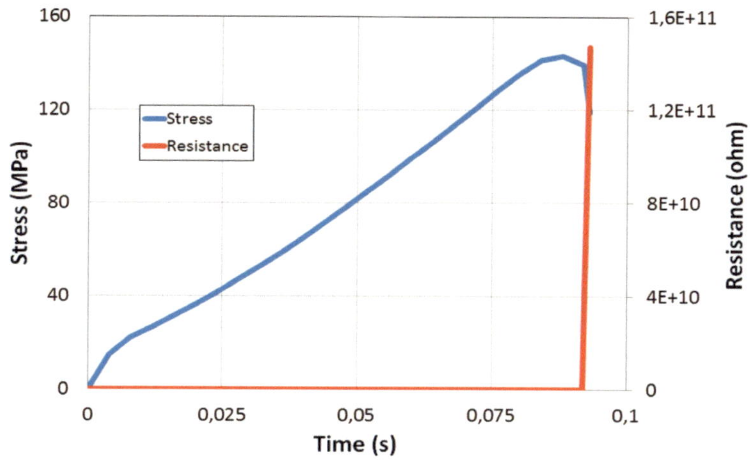

Figure 10.
FE analysis of electromechanical response of Ag-coated monofilament.

7.5 Experimental results of instrumented composite specimen with nylon/silver sensor

Three tensile tests were performed with the composite specimen incorporated with sensor wire. The mechanical response including Young's modulus and yield strength of the tested samples showed that it was not affected by insertion of sensor in the sample and the sensor did not act as intrusive element. The mechanical response of composite specimen and electrical response of the sensor were correlated simultaneously. The resistance was changed at the same time as the failure started to initiate, and, as the test progressed, at the point of failure, the resistance of the sensor started to increase and eventually went to infinity, **Figure 12**. The sensor was reporting what was going on as the fracture formed. The results were

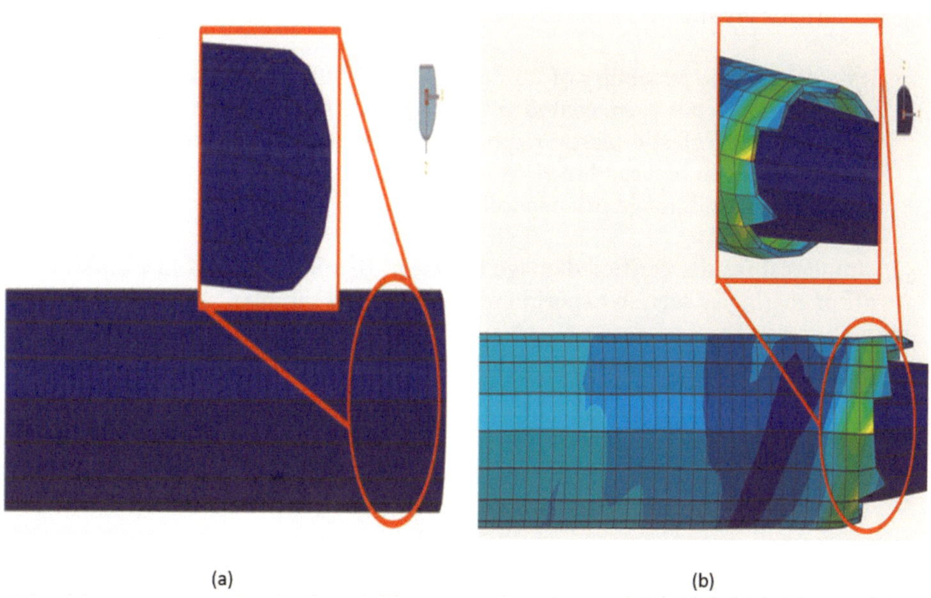

Figure 11.
3D discrete model (a) before failure and (b) after failure.

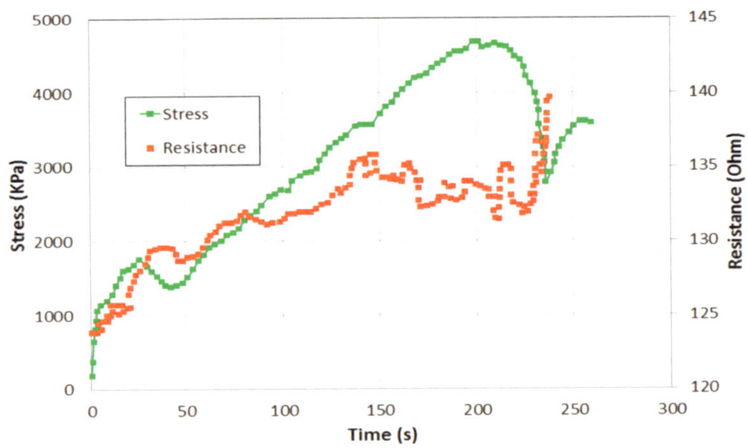

Figure 12.
Electrical response of the sensor with mechanical behavior of composite specimen.

very promising and piezo-resistive flexible sensor was responding very well to the mechanical change in the composite specimen.

8. Conclusions

In this study, a flexible strain sensor was fabricated using the electroplating process. The coating was carried out on nylon yarn, and it appeared thin and uniform. Then, the strain sensor was characterized and tested experimentally, and its behavior verified numerically. The results were very encouraging and reproducible. Next, the yarn sensor was incorporated into a composite material and standardized specimens were prepared and tested. The signal from the sensor was correlated perfectly with the local stress-signal of the composite panel. The following conclusions are drawn from this study:

- It confirmed the feasibility of fabricating a flexible strain sensor. The sensor was made of nylon yarn, coated with silver. The resistivity change was recorded in the elastic-plastic region, which showed that these flexible strain sensor wires can be used for SHM in different structures and can sense deformation or damage prior to failure.

- Numerical results verified damage behavior, that is, ductile failure and flaking off of coating as seen in experimentally fractured samples.

- It was found that the incorporation of these sensor wires would not affect the structural integrity of the specimen.

Author details

Yumna Qureshi[1], Mostapha Tarfaoui[1,2*], Khalil K. Lafdi[3] and Khalid Lafdi[2]

1 ENSTA Bretagne, IRDL-FRE CNRS 3744, Brest, France

2 University of Dayton, Dayton, OH, United States

3 Ohio Connections Academy, Mason, Ohio, United States

*Address all correspondence to: mostapha.tarfaoui@ensta-bretagne.fr

References

[1] Campbell CF. Structural Composite Materials. Ohio: ASM International; 2010

[2] El Moumen A, Tarfaoui M, Lafdi K, Benyahia H. Dynamic properties of carbon nanotubes reinforced carbon fibers/epoxy textile composites under low velocity impact. Composites Part B: Engineering. 2017;**125**(1–8)

[3] Tarfaoui M. Experimental investigation of dynamic compression and damage kinetics of glass/epoxy laminated composites under high strain rate compression. Advances in Composite Materials-Ecodesign and Analysis. 2011

[4] Nachtane M, Tarfaoui M, El Moumen A, Saifaoui D. Damage prediction of horizontal axis marine current turbines under hydrodynamic, hydrostatic and impacts loads. Composite Structures. 2017;**170**:146-157

[5] Shah OR, Tarfaoui M. Determination of mode I & II strain energy release rates in composite foam core sandwiches: An experimental study of the composite foam core interfacial fracture resistance. Composites Part B: Engineering. 2017;**111**:134-142

[6] El Moumen A, Tarfaoui M, Hassoon O, Lafdi K, Benyahia H, Nachtane M. Experimental study and numerical modelling of low velocity impact on laminated composite reinforced with thin film made of carbon nanotubes. Applied Composite Materials. 2018;**25**(2):309-320

[7] ASM Handbook, Volume 21: Composites. ASM International; 2001

[8] Awad ZK, Aravinthan T, Zhug Y, Gonzalez F. A review of optimization techniques used in the design of fibre composite structures for civil engineering applications. Materials and Design. 2012;**33**:534-544

[9] Tarfaoui M, El Moumen A, Lafdi K. Progressive damage modeling in carbon fibers/carbon nanotubes reinforced polymer composites. Composites Part B: Engineering. 2017;**112**:185-195

[10] Arbaoui J, Tarfaoui M, Bouery C, El Malki Alaoui A. Comparative study of mechanical properties and damage kinetics of two- and three-dimensional woven composites under high-strain rate dynamic compressive loading. International Journal of Damage Mechanics. 2016;**25**(6):878-899

[11] Arbaoui J, Tarfaoui M, Arbaoui J, El Malki Alaoui A. Mechanical behavior and damage kinetics of woven E-glass/vinylester laminate composites under high strain rate dynamic compressive loading: Experimental and numerical investigation. International Journal of Impact Engineering. 2016;**87**:44-54

[12] Ihn JB, Chang FK. Pitch-catch active sensing methods in structural health monitoring for aircraft structures. Structural Health Monitoring. 2008;7 (1):5-19

[13] Lynch JP, Law KH, Kiremidjian AS, Kenny TW, Carryer E, Partridge A. The design of a wireless sensing unit for structural health monitoring. In: 3rd International Workshop on Structural Health Monitoring; Stanford, CA; 2001

[14] Yuan FG. Structural Health Monitoring (SHM) in Aerospace Structures. UK: Oxford, Elsevier; 2016

[15] Michaels JE. Detection, localization and characterization of damage in plates with an in situ array of spatially distributed ultrasonic sensors. Smart Materials and Structures. 2008;**17**(3)

[16] Zhu XP, Rizzo P, Marzani A, Bruck J. Ultrasonic guided waves for nondestructive evaluation/structural health monitoring of trusses. Measurement Science and Technology. 2010;**21**(4)

[17] Sassi S, Tarfaoui M, Ben Yahia H. In-situ heat dissipation monitoring in adhesively bonded composite joints under dynamic compression loading using SHPB. Composites Part B: Engineering. 2018;**54**:64-76

[18] Tarfaoui M, El Moumen A, Ben Yahia H. Damage detection versus heat dissipation in E-glass/epoxy laminated composites under dynamic compression at high strain rate. Composite Structures. 2018;**186**:50-61. DOI: 10.1016/j.compstruct.2017.11.083

[19] Loayssa A. Optical fiber sensors for structural health monitoring. In: Mukhopadhyay SC, editor. New Developments in Sensing Technology for Structural Health Monitoring. Berlin, Heidelberg; 2011. pp. 335-358

[20] Lin B, Giurgiutiu V. Modeling and testing of PZT and PVDF piezoelectric wafer active sensors. Smart Materials and Structures. 2006;**15**(4):1085-1093

[21] Raghavan A, Cesnik CES. Review of guided-wave structural health monitoring. The Shock and Vibration Digest. 2007;**39**(2):91-114

[22] Zilberstein V, Walrath K, Grundy D, Schlicker D, Goldfine N, Abramovici E, et al. MWM eddy-current arrays for crack initiation and growth monitoring. International Journal of Fatigue. 2003; **25**(9–11):1147-1155

[23] Speckmann H, Henrich R. Structural health monitoring (SHM)—Overview on technologies under development. In: Proceedings of 16th WCNDT; 2004

[24] Christian B. Next generation structural health monitoring and its integration into aircraft design. International Journal of Systems Science. 2000;**31**(11):1333-1349

[25] Varadan VK, Varadan V. Microsensors, microelectromechanical systems (mems), and electronics for smart structures and systems. Smart Materials and Structures. 2000;**9**(6): 953-972

[26] Trifigny N, Kelly FM, Cochrane C, Boussu F, Koncar V, Soulat D. PEDOT: PSS-based piezo-resistive sensors applied to reinforcement glass fibres for in situ measurement during the composite material weaving process. Sensors. 2013;**13**(8):10749-10764

[27] Atalay O, Kennon WR. Knitted strain sensors: Impact of design parameters on sensing properties. Sensors. 2014;**14**(3):4712-4730

[28] Bashir T. Conjugated polymer-based conductive fibers for smart textile applications [Ph.D. Dissertation]. Sweden: Department of Chemical Engineering and Biotechnology, Chalmers University; 2013

[29] Kaur G, Adhikari R, Cass P, Bown M, Gunatillake P. Electrically conductive polymers and composites for biomedical applications. Royal Society of Chemistry Advances. 2015;**5**:37553-37567

[30] Djaja RG, Moses PJ, Carr AJ, Carnaby GA, Hankoff LD. Finite element modeling of an oriented assembly of continuous fibers. Textile Research Journal. 1992;**62**(8):445-457

[31] Munro WA, Carnaby GA, Carr AJ, Moss PJ. Some textile applications of finite-element analysis. Part I: Finite elements for aligned fibre assemblies. Journal of the Textile Institute. 1997;**88** (4):325-338

[32] Munro WA, Carnaby GA, Carr AJ, Moss PJ. Some textile applications of

finite-element analysis. Part II: Finite elements for yarn mechanics. Journal of the Textile Institute. 1997;**88**(4):339-351

[33] He W, Wang X, Zhang S. Mechanical behavior of irregular fibers. Part II: Nonlinear tensile behavior. Textile Research Journal. 2001;**71**(11): 939-942

[34] Keefe M, Edwards DC, Yang J. Solid modeling of yarn and fiber assemblies. Journal of The Textile Institute. 1992;**83** (2):185-196

[35] Adanur S, Liao T. 3D modeling of textile composite preforms. Composites Part B: Engineering. 1998;**29**(6):787-793

[36] Jiang Y, Chen X. Geometric and algebraic algorithms for modelling yarn in woven fabrics. Journal of the Textile Institute. 2005;**96**(4):237-245

[37] Sriprateep K, Pattiya A. Computer aided geometric modeling of twist fiber. Journal of Computer Science. 2009;**5**(3): 221-225

[38] Vassiliadis S, Kallivretaki A, Provatidis C. Mechanical modelling of multifilament twisted yarns. Fibers and Polymers. 2010;**11**(1):89-96

[39] Peirce FT. Tensile tests for cotton yarns. V.—The weakest link theorems on the strength of long and of composite specimens. Journal of the Textile Institute Transactions. 1925;**17**(7):T355-T368

[40] Hearle JWS, Grosberg P, Backer S. Structural Mechanics of Fibers, Yarns, and Fabrics. New York: John Wiley & Sons Inc; 1969

[41] Treloar LRG, Riding G. Theory of the stress-strain properties of continuous-filament yarns. Journal of the Textile Institute Transactions. 1963; **54**:T156-T170

[42] Riding G, Wilson N. The stress–strain properties of continuous-filament yarns. Journal of the Textile Institute Transactions. 1965;**56**(4):T205-T214

[43] Liu T, Choi KF, Yuan L. Mechanical modeling of singles yarn. Textile Research Journal. 2007;**77**:123-130

[44] Cartraud P, Messager T. Computational homogenization of periodic beam-like structures. International Journal of Solids and Structures. 2006;**43**(3–4):686-696

[45] Lee JH, Kim NR, Kim BJ, Joo YC. Improved mechanical performence of solution-processed MWCNT/Ag nanoparticle composite films with oxygen-pressure-controlled annealing. Carbon. 2012;**50**:98-106